Accession no.
36165169

Mapping

KU-282-160

WITHDRAWN

Critical Introductions to Geography

Critical Introductions to Geography is a series of textbooks for undergraduate courses covering the key geographical subdisciplines and providing broad and introductory treatment with a critical edge. They are designed for the North American and international market and take a lively and engaging approach with a distinct geographical voice that distinguishes them from more traditional and out-dated texts.

Prospective authors interested in the series should contact the series editor:

John Paul Jones III
Department of Geography and Regional Development
University of Arizona
jpjones@email.arizona.edu

Published

Cultural Geography
Don Mitchell

Political Ecology
Paul Robbins

Geographies of Globalization
Andrew Herod

Geographies of Media and Communication
Paul C. Adams

Social Geography
Vincent J. Del Casino Jr

Mapping
Jeremy W. Crampton

Forthcoming

Environment and Society
Paul Robbins, Sarah Moore, and John Hintz

Geographic Thought
Tim Cresswell

Cultural Landscape
Donald Mitchell and Carolyn Breitbach

Research Methods in Geography
Basil Gomez and John Paul Jones III

Mapping

A Critical Introduction to Cartography and GIS

Jeremy W. Crampton

LIS LIBRARY

Date	Fund
02/08/12	g-che

Order No
2325986

University of Chester

⟨W⟩WILEY-BLACKWELL

A John Wiley & Sons, Ltd., Publication

This edition first published 2010
© 2010 Jeremy W. Crampton

Blackwell Publishing was acquired by John Wiley & Sons in February 2007.
Blackwell's publishing program has been merged with Wiley's global Scientific,
Technical, and Medical business to form Wiley-Blackwell.

Registered Office
John Wiley & Sons Ltd, The Atrium, Southern Gate, Chichester, West Sussex, PO19 8SQ,
United Kingdom

Editorial Offices
350 Main Street, Malden, MA 02148-5020, USA
9600 Garsington Road, Oxford, OX4 2DQ, UK
The Atrium, Southern Gate, Chichester, West Sussex, PO19 8SQ, UK

For details of our global editorial offices, for customer services, and for information about
how to apply for permission to reuse the copyright material in this book please see our
website at www.wiley.com/wiley-blackwell.

The right of Jeremy W. Crampton to be identified as the author of this work has been
asserted in accordance with the Copyright, Designs and Patents Act 1988.

All rights reserved. No part of this publication may be reproduced, stored in a retrieval
system, or transmitted, in any form or by any means, electronic, mechanical, photocopying,
recording or otherwise, except as permitted by the UK Copyright, Designs and Patents Act
1988, without the prior permission of the publisher.

Wiley also publishes its books in a variety of electronic formats. Some content that appears
in print may not be available in electronic books.

Designations used by companies to distinguish their products are often claimed as
trademarks. All brand names and product names used in this book are trade names,
service marks, trademarks or registered trademarks of their respective owners. The publisher
is not associated with any product or vendor mentioned in this book. This publication is
designed to provide accurate and authoritative information in regard to the subject matter
covered. It is sold on the understanding that the publisher is not engaged in rendering
professional services. If professional advice or other expert assistance is required,
the services of a competent professional should be sought.

Library of Congress Cataloging-in-Publication Data

Crampton, Jeremy W.
　　Mapping: a critical introduction to cartography and GIS / Jeremy W. Crampton.
　　　　p. cm. – (Critical introductions to geography)
　　Includes bibliographical references and index.
　　ISBN 978-1-4051-2172-9 (hardcover : alk. paper) – ISBN 978-1-4051-2173-6 (pbk. : alk.
paper)　1. Cartography–Computer programs.　2. Geographic information systems.
I. Title.
　　GA102.4.E4C73 2010b
　　526—dc22

2009043996

A catalogue record for this book is available from the British Library.

Set in 10.5/13pt Minion by Graphicraft Limited, Hong Kong
Printed in Singapore by Ho Printing Singapore Pte Ltd

01　2010

Contents

Acknowledgments

I was first asked to do this book several years ago but due to other projects I very luckily turned it down – I can't imagine how relevant this book would be if it had been completed before the radical changes in mapping brought about through Google Earth and the geoweb! So I first thank J. P. Jones and the editorial staff at Blackwell (especially Justin Vaughan and Ben Thatcher) for asking me to reconsider and for being so patient. I first wrote about the Peters world map controversy in the early 1990s, assisted by Francis Herbert of the Royal Geographical Society who found an original copy of Gall's *Easy Guide to the Constellations* at a book stall in London, and helped me access Gall's correspondence at the RGS. For the new material in this book I am grateful to Trevor Gould at the Carrubber's Christian Centre in Edinburgh for previously unpublished internal documents on the Rev. James Gall, and to Bob Abramms of ODT Inc. for a DVD of Arno Peters' last interviews. Additionally, David Livingstone provided me with information on the Preadamites and their relationship to racial discourses. David Woodward and Dalia Veranka both shared with me many of their memories of Brian Harley, and Matthew Edney taught me much about Harley, Woodward, and the history of cartography. I'm glad to count them as my friends. The many spirited conversations I've been privileged to have with Denis Wood over the last 15 years have benefited me enormously. My biggest single intellectual debt is to John Krygier whom I first met in graduate school and who has been a constant source of intellectual provocation ever since.

Geoffrey Martin, AAG archivist and professor emeritus, Southern Connecticut State University, generously opened up his home to me and provided me with many of his privately collected documents (in some cases now the only extant copies after the American Geographical Society destroyed many of its own records). His insider knowledge of Anglo-American geography was recently recognized by the American Geographical History Award from the AAG. Archivists at Yale, Eastern Michigan University, Johns Hopkins, the American Geographical Society Library, and the

National Archives were instrumental in orienting me to their material and maximizing my visits.

Several artists who experiment and work with maps kindly shared their work with me, including Deirdre Kelly, Robert Derr, Nik Schiller, Steven R. Holloway, Matthew Knutzen and kanarinka (Catherine D'Ignazio). Thanks to Ed Dahl for the picture of Brian Harley.

John Pickles' contributions to mapping need no introduction. John has been a good friend and generous colleague. I'd like to thank him, Eric Sheppard, and Tom Poiker for organizing the Friday Harbor meetings in 1993.

Kara Hoover provided me with a backstop against my more egregious anthropological errors. Any that remain are the result of my own shortcomings.

Some parts of Chapters 1 and 2 appeared in an earlier form in *ACME: The Journal for Critical Geographies*. Chapter 6 draws from material previously published in *Cartographica*, while an earlier version of Chapter 8 appeared in the *Geographical Review*. I would like to thank the editors and publishers of these journals for their generosity in allowing me to use this material.

I dedicate this book to my father, William George Crampton (1936–97), who nearly 40 years ago took me on hikes across Offa's Dyke and Hadrian's Wall and taught me how to read an Ordnance Survey map.

J.W.C.
Atlanta
April 2009

Figures

Tables

About the Cover: Size Matters

The cover shows a map of the world's population created by Mark Newman. However, instead of showing the world as we usually see it, either from space or on a projection, Newman has created a map in which the size of each country is directly proportional to its population. This map, known to cartographers as a cartogram, offers a radical reinterpretation of the familiar world, as if the landmasses have been created by a malfunctioning lava lamp. China and India – which alone have about one-third of the world's population – are of course huge, but so also are Indonesia and Nigeria. The United Nations predicts that by 2050 Nigeria will be the fifth biggest country in the world, up from 15th in 1950. Countries such as Canada, Australia, and Russia, which tend to dominate most maps we normally see, are here revealed for what they are: relatively lightly populated. Interestingly South America comes through relatively unscathed and is the closest to "normal" maps. The blue line representing the equator also reveals another rather startling fact: the "global North" is by far the more highly populated half of the planet; the hemispheres are far from equal. The map is also good at suggesting reasons for regional geopolitics: look at the sizes of the Ukraine, Turkey, or Ethiopia for example.

Cartograms can be made from many kinds of data, and Newman and his colleague Danny Dorling have produced maps of people living with HIV/Aids, spending on healthcare, GDP, CO_2 emissions, and maps of the US election by number of voters.

Finally, it's irresistible to compare the cartogram to the playful surrealist map in Chapter 2. On what basis is *that* map drawn?

Chapter 1

Maps – A Perverse Sense of the Unseemly

This book is an introduction to critical cartography and GIS. As such, it is neither a textbook nor a software manual. My purpose is to discuss various aspects of mapping theory and practice, from critical social theory to some of the most interesting new mapping practices such as map hacking and the geospatial web. It is an appreciation of a more critical cartography and GIS.

Why is such a book needed? We can begin with silence. If you open any of today's prominent textbooks on cultural, political, or social geography it is more than likely that you will find little or no discussion of mapping, cartography, or GIS. A recent and well-received book on political geography for example (Jones et al. 2004) makes no mention of maps in any form, although it is subtitled "space, place and politics." Similarly, Don Mitchell's influential book on cultural geography and the precursor to the series in which this book appears (Mitchell 2000) deals at length with landscape, representation, racial and national geographies, but is completely and utterly devoid of the role of mapping in these important issues (and this despite Mitchell's call for a "new" cultural geography that does not separate culture from politics!). And while a book on *Key Concepts in Geography* (Holloway et al. 2003) can state that "geographers have . . . studied the ways in which maps have been produced and used not only as objects of imperial power but also of postcolonial resistance" (Holloway et al. 2003: 79) the subject is then quietly dropped. Yet is it the fault of these authors – accomplished scholars – that maps and mappings are not considered part of larger geographical enquiry?

For there is a second silence. Cartographers and GIS practitioners themselves have had very little to say about politics, power, discourse, postcolonial resistance, and the other topics that fascinate large swaths of geography and the social sciences. Open any cartography or GIS textbook and you will find only deep silence about these matters. There are few cartographic voices examining the effects of GIS and mapping in the pursuit of homeland security. There are no journals of cultural or political cartography. What percentage of GIS applications are being created to address poverty? Is there feminist mapping? And if GIS and mapping have always coexisted alongside military and corporate applications then how many GIS practitioners have critically analyzed these relationships? Perhaps most arresting is the increasing separation of GIS and mapping from geography as a whole. In other words the evolution of GIScience as a technology-based subject rather than a geographic methodology (for example the focus in GIScience on formal "ontologies"). In sum, one might be forced to conclude that mapping is either *incapable* of such concerns, or that it rejects them.

This book is an introduction to these questions, and in part an answer to them from a critical perspective. It is an attempt to push back against the common perception that cartography and GIS are not concerned with geographical issues such as those listed above. The basic viewpoint is that mapping (i.e., cartography and GIS) is both capable of engaging with critical issues, and has often done so. While the word "critical" may be overused and ironically is itself in danger of being used uncritically (Blomley 2006), I believe its application to mapping remains fruitful and exciting. And rather than some trendy new term, there is a long and remarkable critical tradition in cartography and GIS, if in a "minor" and subjugated way. If it did not appear full-blown on the scene in the late 1980s (as the story usually goes, see Chapter 4) the critical traditions in cartography (often accessible through a historical genealogy) demonstrate how mapping and the wider field of geographical enquiry worked together for many years.

If you look back at the history of mapping it might appear that to be a "cartographer" meant to be a mapmaker, someone whose profession it was to draw maps (the word "cartography" is of early nineteenth-century origin, but "map" has a much longer history, see Krogt [2006]). It was only in the twentieth century that one could be a cartographer who *studied* maps but didn't necessarily make them (or have any skill in making them) – that is, with the development of the discipline of cartography as a field of knowledge and enquiry. In this sense the discipline of cartography started to become divorced from its practice in the sense of map production. This might seem rather unusual. After all, there are no geographers doing geography and then a bunch of people in academia who study them and how they work! To be a geographer (or a physicist or chemist) is to *do* geography, physics, or chemistry.

But this initial distinction between mapmaking and cartography as discipline is quite hard to maintain. Although "mapmaking" in the traditional sense – as Christopher Columbus might have practiced it for example – with all of its pens, paper sheets, sextants, watermarks, and mastery of hand-drawn projections obviously has very little role in academic study today, you will nevertheless still find yourself doing mapping. Except you might call it GIS, geomatics, surveying, real-estate planning, city planning, geostatistics, political geography, geovisualization, climatology, archaeology, history, map mashups, and even on occasion biology and psychology. And in geography too we could probably agree that there are a bunch of people "doing" human geography who are distinguishable (sometimes) from the academics studying them. Just think of all those articles on the Research Assessment Exercise

(RAE) or on which journals geographers publish in. And finally there are the objects of critique, the (im)material products and processes of mapping and GIS. All three of these; objects, do-ers or performers of mapping, and the production of critique have complex interrelationships.

The point then is not that long ago there was something called mapmaking (which is now called geospatial technology or GIS) but rather that the understanding of what people thought they were doing with things they called maps has changed over time, as well as over space.

One of the stories that I was taught as a student is that cartography became scientific only recently, say after World War II. It did so, the story went, largely for two reasons. First, it finally threw off art and subjectivity (here reference was often made to the work of Arthur Robinson and his call for formal procedures of map design). Thus science was posed in opposition to art. Second, it became as it were "post-political" by throwing off the fatal attraction to propaganda and ideological mapping evidenced prior to and during the war, and promoting a kind of Swiss-like neutrality about politics. In doing so it paralleled the path taken by the discipline of political geography, which also found itself tarnished by its cooption during the war. But where political geography went into decline until the 1970s (Brian Berry famously called it a "moribund backwater" [R. Johnston 2001]), cartography tried to insulate itself from politics altogether by gathering around itself the trappings of objective science. The map does exactly what it says on the tin.

Yet both of these developments are myths. As the critical work of writers such as Matthew Sparke, Denis Cosgrove, and Anne Godlewska has shown, mapping as a discipline and as a practice failed to establish a rigid separation from art, nor did it ever become post-political. Chapters 5 and 12 document these myths in more detail and show what the critical response has been.

In a recent provocative article Denis Wood issued a heartfelt cry that "cartography is dead (thank God!)" (D. Wood 2003). By this he meant that the gatekeepers, academic cartographers, dwelling as it were like a parasite on actual mapping, were dying off. Maps themselves, meanwhile, have never been healthier – if only disciplinary academics would leave them be! While I have some sympathies for this position (who wants gatekeepers except other gatekeepers?) I'm not quite sure it's correct. Rather, first because the study of mapping continues as never before, GIS is something like a $10 billion a year corporate-military business, and the advent of map hacking and map mashups has released the inner cartographer in millions of ordinary people. And second, I'm not sure it's possible to separate mapping practice from mapping discourses quite so neatly (that minor critical tradition again!). In fact practices and discourses are intimately intertwined.

Not that discourses or knowledge go uncontested. If it was when cartography became formalized as a discipline that mapping was valorized as "scientific," then by the 1990s a number of geographers, cartographers, and GIS practitioners drew on the larger intellectual landscape to renew a critical spirit. Today we are still drawing on that renewed linkage between mapping and geography. The central rationale of this book therefore is to demonstrate the relevance of spatial knowledge production in GIS and cartography as critical for geographers, anthropologists, sociologists, historians, philosophers, and environmental scientists.

Yet it is also plain to see that mapping has undergone a tremendous re-evaluation over the last 15 years (or longer). In accounts of this period (Schuurman 2000; Sheppard 2005), the story is told of how the encounters between mapping and its critics began with mutual suspicion and ended up with something like mutual respect. Sheppard further argues that what began with investigations of the mutual influences between GIS and society has become a "critical" GIS (with "GIS and society" representing the past and critical GIS representing the future). By this he means not just a questioning approach, but one that is critical in the sense used in the wider fields of geography and critical theory. This sense includes Marxist, feminist, and post-structural approaches among others. For Sheppard critique is a "relentless reflexivity" which problematizes various power relationships.

This narrative can itself be problematized by showing that beneath the official histories of GIS and mapping lie a whole series of "counter-conducts." These dissenting voices, sometimes speaking past one another, sometimes speaking out from below, are discussed in more detail in Chapter 2. There is therefore a minor as well as a major history of mapping and GIS, a series of "subjugated knowledges" (Foucault 2003b) that while they have popped up from time to time in the past are now making themselves felt as never before. In particular I think it is fruitful to see the history of critical GIS and cartography not as something that has only recently occurred, but one that in fact can be seen at other more distant times as well. This is what Foucault means by subjugated knowledges; ones that for whatever reason did not rise to the top, or were disqualified (for example, for not being scientific enough). But it doesn't mean they weren't there. Furthermore, Foucault suggests that it is the reappearance of these local knowledges alongside the official grand narratives that actually allows critique to take place. This is also an idea that we shall examine in the next chapter.

This book then appears at a transitional moment in the history of GIS and mapping. Great changes are occurring and it would be wrong to say we know exactly where they are leading. The following diagram summarizes some of these tensions which are fluctuating throughout mapping. This diagram is meant to be indicative rather than complete. Imagine that the space transected by the tensional vectors is a rubber sheet being stretched out (readers with multi-dimensional imaginations could also see it as an expanding sphere). As the sheet is stretched the field gets larger – but also thinner, perhaps dangerously so in some places (Figure 1.1).

This figure illustrates how mapping is a field of power/knowledge relations being simultaneously taken in different directions. On one axis, critical approaches, with their "one–two punch" of theoretical critique (Kitchin and Dodge 2007) and the emergence of the geoweb are questioning expert-based mapping. The increasing use of mapping technologies among so-called amateurs or novices (for example the 350–400 million downloads of Google Earth) is reshaping all sorts of new spatial media, and is allowing the pursuit of alternative knowledges. Meanwhile, on another

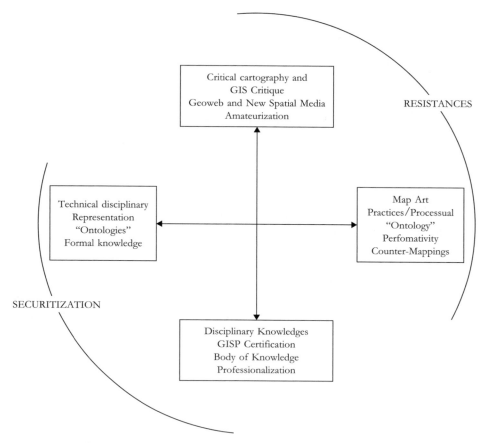

Figure 1.1   The field of tension in mapping.

axis, there are very real trends toward nailing knowledge down into a coherent "body" that can be mastered by experts. We'll know they are experts because they hold a certification. What we're talking about here then is a clerisy or set of experts.

The desire for mapping to be post-political is exemplified in the diagram by those who focus on the technical issues in isolation from their larger socio-political context. Many cartography and GIS journals have now become almost completely dominated by technical issues, research which no doubt reflects the research agendas pursued by the next generation of PhDs – of which you may be one.

These different directions can be broadly described as a trend toward "securitization" of knowledge in the one direction and "resistances" in the other. Securitization of information refers to the efforts that are made to anchor, control, and discipline geographical knowledges. Another example is the increasing interest among GIScientists in "ontologies" defined as formal, abstract, and computer-tractable definitions of real-world entities and their properties. Certainly it is nothing new to observe that there is a danger whenever technology is involved of taking up mapping *only* as a technology. As the German philosopher Martin Heidegger remarked six decades ago "the essence of technology is by no means anything technical" (Heidegger 1977: 4). But because it is often ignored, the implications of this seemingly counter-intuitive claim are taken up in various ways throughout the book.

The Need for Critique

Why is critique needed? "Critical" approaches to both GIS and cartography play important roles, but are not yet mainstream. It's possible you might feel both that maps are terribly old-fashioned (something you studied in lower school) and yet tremendously exciting (Google Earth and homemade mapping applications, geovisualization, or perhaps human geosurveillance). Where does the truth lie?

Some of these mixed feelings were the topic of discussion in a recent issue of *Area*, one of the UK's better known geography journals. Here's Joe Painter awkwardly confessing that he's in love:

> I love maps. There, I've said it. I am coming out as a cartophile. Although I became fascinated by maps when I was a child (and even once told a school careers advisor that I wanted to work for the Ordnance Survey – Britain's national mapping agency), maps have figured little in my work as an academic geographer. I suspect that many human geographers who learned their trade in the postpositivist 1980s, as I did, shared my mild embarrassment about maps. (Painter 2006: 345)

So Painter may be in love, but it's a love that dare not speak its name: maps figure little in his work. Painter's "cartographic anxiety" (Gregory 1994; Painter 2008) resonates with many people interested in maps and mappings. As the geographer-phenomenologist John Pickles has written, there's a perverse sense of the unseemly about maps (Pickles 2006). These wretched unreconstructed things seem to work so *unreasonably well*! This sense of mapping as unseemly and unwelcome is often assumed as a given by a surprisingly large segment of people. We're ambivalent. In the eyes of critical geographers the success of maps has not come without a price. Haven't maps after all provided the mechanism through which colonial projects have been enabled (Akerman 2009; Edney 1997)? Isn't there a long history of racist mapping (Winlow 2006)? Today, isn't it simply the case that GIS and GPS are essential elements of war (N. Smith 1992)? Wasn't Arthur "dean of modern cartography" Robinson an instrumental part of the Office of Strategic Services – the precursor to the CIA? At the very least, GIS is surely a Trojan horse (Sheppard 2005) for a return to positivism (Pickles 1991)?

These observations are valid. And yet, the same points could be made with reference to geography (or other disciplines such as anthropology) as a whole. Weren't they involved in colonialist projects? Doesn't the past of geography, anthropology, or biology contain racist writing and racist people? Sure.

Item: Madison Grant, who wrote the racist book *The Passing of the Great Race* (guess which race he feared was passing away), was a longstanding Council Member of the American Geographical Society (AGS) including during the time when Isaiah Bowman was Director, and agitated for quota-based laws in the 1920; he also published a version of the book in the AGS journal *Geographical Review* during wartime (Grant 1916).

Item: former President of the Association of American Geographers, Robert DeCourcy Ward, professor of geography at Harvard University, wrote a series of frankly racist eugenicist papers bitterly complaining about the low quality of immigrants into this country (Ward 1922a; 1922b). To influence anti-immigration laws he founded the "Immigration Restriction League" which succeeded in getting a literacy test into the Immigration Act of 1917.

Rather than drawing a veil over these facts, or saying that mapping is essentially a racist or capitalist tool, any honest intellectual history will seek to examine them – not least because their arguments are still reprised today. For instance, biological race is being reinscribed in genetics (Duster 2005) and advocates of English as the official language of the US are still active (30 US states and at least 19 cities have adopted English as their official language).

The first response to "why critique?" is that it is not *un*reason or something fundamentally unknowable that is at stake here, but rather the need to examine the very rationality that animates mapping and GIS today. Not only can this rationality be explained, but it can also be challenged, and it is the job of a critical GIS and critical cartography to do just that.

A second question revolves around the historical complicity of mapping and GIS in military, colonial, racist, and discriminatory practices. It is tempting to see maps and GIS as "essentially" complicit and best avoided. Maps are "nothing more" than tools of capitalist expansion and exploitation. (Sometimes one suspects that this tactic explains the silence of those critical geographers we began with. Maps and GIS are embarrassing!)

One popular response is to deny that maps and GIS are "essentially" anything in particular. Maps and GIS are "neutral" technologies that can be used for both good and bad purposes (whatever they are!). On this view we might readily acknowledge the complicity of geography in colonial projects but also point out that maps and GIS can be used to track organ donations, manage global air travel, and empower local communities to fight off Wal-Mart. They are a little like technologies sitting on a shelf, waiting to be pulled down and used. A direct analogy comes to mind here: atomic power. It can be used to build atomic bombs or to power the national grid. We might argue that each one should be judged on its own merits. The same thing goes for mapping, we might then argue. Sometimes maps are used for bad purposes but sometimes they are used for good ones. Those that are, on the whole, bad, we could criticize. Those that seem to be positive we could praise. This results in an *economy of morality*; a balance between good and bad and which one outweighs the other.

If we take this view, it has the merit of being very flexible. We could assess a range of geosurveillance techniques with it for example (Monmonier 2002b). Monmonier resists the involvement of cartography in anything that might seem "political." Not surprisingly, when I once asked him about the possibility of there being such a thing as a "political cartographer" in an interview, he replied that he thought it a "glib phrase" and that he "would apply the label political cartographer [only] to people who draw election-district maps" (Monmonier 2002a).

Too often however, this response is just a cover for a do-nothing approach. It is a way of fobbing off the power commitment we make whenever we assert or produce knowledge. This view that maps are politically neutral was recently put forward by the influential National Academies of Science (NAS) in a report entitled "Beyond Mapping" (Committee on Beyond Mapping 2006). The committee was comprised of well-known scholars in GIS, cartography, and geography including Joel Morrison (Chair), Michael Goodchild, and David Unwin. The committee was not unaware of the need to examine the societal implications of GIS and mapping as technologies. For example, they write:

> Geographic information systems and geographic information science appear to be benign technologies, but some of their applications have been questioned; as is true of any technology, *GIS, though neutral in and of itself*, can be used for pernicious ends. (Committee on Beyond Mapping 2006: 47, emphasis added)

A critical approach would argue that this appeal to the neutrality of mapping knowledges is a failure.

The things I have been talking about so far constitute some of the aspects of what Derek Gregory once called "cartographic anxiety" (Gregory 1994). His influential little phrase captures how people sometimes seem to feel about maps. An anxiety is a disorder, and if pronounced enough becomes a subject for psychological investigation – a clinical case. It's a contrary and split kind of anxiety (a schizophrenic anxiety?), because on the one hand we have maps involved in those colonialist projects, to de- and resubjectify people, or perhaps in which powerful cartographic imagery is invoked to justify an "axis of evil" (Gregory's more recent work is sustained by a sense of moral outrage at Abu Ghraib and Guantanamo Bay [Gregory 2004; Gregory and Pred 2007]). So we're anxious about using any such uncritical devices that work all too well to establish concrete realities – the *unseemly* perversity, in the sense of the word as unwanted, not in good taste, out of place.

And then there's the other kind of anxiety that Gregory talks about; the anxiety of uncertainty which maps and geographical knowledges produce when their authority is undermined. Citing Gunnar Olsson and Brian Harley's work on "deconstructing" the map (see Chapter 7) Gregory talked of an anxiety (or we might say it is the *perversity*) that arises when knowledge is destabilized, though he was quick to say it did not mean a descent into "giddy relativism" (can we ask why not?) (Gregory 1994: 73).

So now this anxiety has two contradictory parts – on the one hand maps are incredibly powerful devices for creating knowledge and trapping people within their cool gleaming grid lines, on the other they seem to be nothing at all, just mere bits of fluff in the air.[1] Maps are sovereign; maps are dead.

The Third Way?

One might register a few problems with both of these viewpoints however. It is noticeable that the second viewpoint, that of technology being non-essential or "neutral," often crops up when a new technology appears and people are thinking about it for the first time. It's as if people want to try and get things straight in their mind and that this can be done by considering each application "before" or outside of untoward influence. Bringing in politics only serves to muddy the waters.

The problem with these ideas is that they miss the point. Even casting a cursory glance at the history of cartography should lead us to suspect that mapping and maps have a whole series of engagements in politics, propaganda, crime and public health, imperialist boundary-making, community activism, the nation-state, cyberspace, and the internet. That is, mapping has a politics. It is hard to imagine mapping that does not in some way or other involve politics, mapping is itself a political act.

As a politics of mapping, critical cartography and GIS question what kinds of people and objects are formed through mapping. As the Canadian philosopher Ian Hacking puts it, how are people made up (Hacking 2002)? This is a question about how categories of knowledge are derived and applied, a question as old as Kant and as contemporary as racism.

Maps produce knowledge in specific ways and with specific categories that then have effects (i.e., they deploy power). Categories are useful, but at the same time they encourage some ways of being and not others. Often, some ways of being are accepted as somehow typical and are called "normal," while others are called "abnormal." Then there is a tendency to try and correct, eliminate, or manage the abnormal.

Maps and mapping are not the only rationalities at work in society, but it is interesting that the maps we commonly find in modern-day GIS (i.e., thematic maps) were all invented around the same time: the early nineteenth century (Robinson 1982). This was the time that another set of great techniques were developed that we increasingly rely on, namely statistics and the emergence of the theory of probability (Hacking 1975). Both maps and statistics were two great technologies of management that are used extensively by governments to get a grip on risk and threats to the country. The most recent demonstration of this was the aftermath of the terrorist attacks of September 11, 2001. Maps and GIS were deployed to analyze "at-risk" targets, or to surveille "risky" populations. For example, the FBI deemed that mosques in the USA were risk factors, and shortly after 9/11 began building a database of all the mosques in America (Isikoff 2003).

What is the result of such surveillance and what kinds of people does it "make up"? This is a different kind of question than one that weighs the plusses and minuses of technology. It is one focused on power, discourse, politics, and knowledge. These are the questions that critical cartography and GIS is interested in: the "third way" between saying that we should examine the *essential* nature of mapping, and saying that mapping is empty or neutral. It would examine instead the way that maps and GIS are situated in specific times and places, what knowledges they produced, and with what effects.

When this third way of critical mapping started to become more noticeable about 15 or 20 years ago, maps and mapping were only studied by cartographers. When GIS started becoming popular around the same time (the early 1990s) it was no longer possible to see mapping as a minority interest, if only because a number of well-placed geographers raised serious objections to it. Perhaps the most famous was the comment in the late 1980s by the then-president of the American Association of Geographers (AAG) that GIS did not belong in the "intellectual core" of the discipline, being merely a technique (Jordan 1988). These comments did little to endear critics to GIS users and vice versa (Sheppard 1995; 2005). This series of dissenting voices, sometimes speaking past one another, sometimes speaking out from below are taken up in Chapters 2 and 4.

Some of these voices are well known, such as that of Arno Peters who launched an attack on cartography for its complicity in racist geographies (and who in return was vividly counter-attacked by virtually all cartographers). Other voices are less well known. Who today remembers J. Paul Goode as anything but the successful author of the standard college atlas? Yet he railed against the "evil Mercator" projection in terms remarkably similar to those of Peters (Chapter 7). Other voices are coming from outside academia entirely. The phenomena of map-blogs and the "geospatial web" operating for the most part well outside the view of academia raises the question of where and how innovation is occurring in mapping today. Is it occurring within the discipline at all? If not, what does that mean about the quality of mapping – and the future of the discipline? Is there a new populist "peasantry" on the march (Chapter 3)?

In denying a relationship between mapping and politics, cartography and GIS have evidenced similar intellectual histories as other technological fields that generate knowledge (Misa et al. 2003). But if knowledge can be generated in these technical fields, then that knowledge is always put into play, as it were, in competition, with some knowledges succeeding (especially those with a scientific orientation) and others being relegated. So again, knowledge is related to power. Some knowledge is easy to obtain, while some, if it is not actively suppressed, is marginalized and ignored. Native or indigenous cartographic knowledge, for example, is a case in point. Until fairly recently very little was known about non-Western cartographies because they were not easy to reconcile with a story of cartography as an ever-more accurate and scientific representation of the landscape. Against this, Edney has called for a "history [of cartography] without progress" that would recognize the backtracking, "wrong" turns, and diversions even within the Western tradition (Edney 1993). Meanwhile the rich tradition of indigenous mapping that operates independently of such terms as progress and science is attracting renewed attention (Sparke 1995; 1998; Turnbull 1993; Woodward et al. 2001).

A second disciplining process can be traced to the period following World War II, when modern cartography came of age. As I discuss in Chapter 5, during the war a number of American scholars led by Arthur Robinson worked to draw cartography together into a discipline. As they did so however, they created a certain view of mapping (and, by extension, GIS) that shied away from any involvement of maps with political issues. Perhaps understandably, these writers took from their wartime experiences the need to avoid the excesses of propaganda that had infected cartography and had caused what Pickles calls a "crisis of representation" (Pickles 2004). Instead of propaganda, maps should be used to tell the truth as clearly as possible, within the limits of the map form. This meant not only paying attention to map design (a field more or less invented only after the war, but which drew heavily from graphic design), but also to the way that maps were used by actual people, or in other words the field of map user studies. This move took cartography away from politics in much the same way that political geography also shied away from politics in the same period and for essentially the same reasons (political geography was at the time called a "moribund backwater" by Brian Berry [quoted in Agnew 2002: 17]).

These two reasons then – the positioning of cartography as a technological or scientific field, and the post-war move away from socio-political issues – have, I would suggest, served to isolate cartography from the wider discipline of geography.

A Note on Terminology

Over the years much has been written examining the relationship between two fields of practice: cartography and GIS. I remember at the 1996 meeting of the Association of American Geographers (AAG) the then-president, Judy Olsen (a cartographer), held a Presidential Plenary session on the question "has GIS killed cartography?" This reflected a fear in the cartographic community that GIS would be the end of cartography (either as a discipline or as an employment option). Now, more than ten years later, it would appear that many of those fears have come to pass, but in a somewhat contradictory way. The job market is certainly one which speaks of GIS and geospatial information. But instead of being "killed," mapping transformed itself, firstly by emphasizing itself as "geographic visualization" in the 1990s, and secondly through its role in map hacking and the geospatial web (see Chapter 3 [and D. Wood 2003]). And it turned out that GIS was most often used to make maps anyway, and is a lot less quantitative and more qualitative than some people might think (Kwan and Ding 2008; Pavlovskaya 2006).[2] Ironically, now it is GIS that is playing catch-up, as the public flocks to software such as Google Earth and map mashups (Erle et al. 2005). The use of maps and mapping tools (and hence if you like the number of cartographers – amateur as well as professional) has never been higher.

This perspective has a number of advantages. It allows us to focus on the question, what is mapping, today? This in turn allows us to cut short any attempts to say, once and for all, what mapping, cartography or GIS essentially are, as if they existed outside time and to divide this portion into something called cartography and this portion into GIS. For this reason I will not (perhaps surprisingly) offer an answer to the question "what is a map?" (Vasiliev et al. 1990). Where in a recent article Martin Dodge and Rob Kitchen try to get the reader to answer this question at the ontological level (Kitchin and Dodge 2007) I will speak of maps as historically situated practices and discourses. That is, I will lay my cards out by admitting that what interests me the most is epistemology or knowledge: its creation, its relation to power and politics; in sum its effects on people and places.

This opens up a whole historical perspective. For anyone who, like me, was dulled into submission by high school history, it is something of a liberation – and something of a shock – to realize that history is not dead, gone, and useless, but something which actively shapes us here and now in the present. There are good reasons then for what we are now, and, through an intellectual history, these reasons can be traced out, with a view perhaps to breaking the grip of that past and of creating something new.

In order to make some headway in all this I will make the claim that it is not so much the specific technology that should concern us, but rather the "mapping tradition" that exists in any given moment. I define "mapping" deliberately loosely as a human activity that seeks to make sense of the geographic world, it is a way in which we "find our way in the world" (Crampton 2003). What are the possible ways of knowing, geographically? Whether this is via the dreamtime maps of native Australians, the latest release of GIS software, or handheld devices that audibly announce our location to us matters less than that human yearning for understanding. In this book I shall therefore use the term mapping to refer to *both* cartography and GIS because despite their differences I believe they are both part of that tradition, a tradition that stretches back to the earliest recorded human history – and even longer (Smith 1987). Furthermore, I shall agree with Clarke (2003) that GIS has its roots in cartography and is in that sense the way that mapping is practiced today (GIS was developed as a technology in the 1960s and as a science – GIScience – in the 1990s [Goodchild 1992]). I know from my contributions to the GIS trade journal *GeoWorld* there are many professional GIS users who would disagree with this assertion and who believe that today cartography is part of GIS, not the other way round. But I think that this both ignores the historical dimension of the mapping tradition, and gives undue emphasis to cartography and GIS as chiefly technical endeavors. At bottom, we deploy both maps and GIS analysis because we want to make sense of the geographical world.

Notes

1   Cf. Communist Manifesto, "all that is solid melts into air."
2   I once this mentioned to a geology colleague who was insisting that GIS could only be quantitative. "Maybe I have misunderstood what GIS is then," he said, puzzled. To which the response would be – yes.

Chapter 2

What Is Critique?

Introduction

This book is a *critical* introduction to mapping and GIS. It is part of a series of books that provide critical introductions to various aspects of geography. Looking around, we find interest in critical geography has blossomed over the last ten or fifteen years, with books, articles, conferences, and even an online discussion list. So what *is* critique? Where does it come from? Is everyone critical now and if so how is it different from "uncritical" geography?

First, we might clear a common misunderstanding. A critique is not a project of finding fault, but an examination of the assumptions of a field of knowledge. Its purpose is to understand and suggest alternatives to the *categories of knowledge* that we use. Michel Foucault, who often worked in a spirit of critique, put it this way:

> A critique does not consist in saying things aren't good the way they are. It consists in seeing on what type of assumptions, of familiar notions, of established, unexamined ways of thinking the accepted practices are based. (Foucault 2000c: 456)

These "unexamined ways of thinking" (i.e., assumptions and familiar notions) shape our knowledge. For example, in cartography textbooks it is often assumed that good map design must achieve "figure-ground" separation (the separation of the main object from its background), even though recent research on cultural differences in the perception of figure-ground reveals that non-Western viewers do not have the same reaction to figure-ground as Western viewers (Chua et al. 2005). For example, in Figure 2.1 many readers with Western backgrounds will be able to make two different interpretations, one in the foreground and one in the background.

Chua et al. discovered that due to cultural differences, people from Europe and Asia will "see" a scene in different ways. For over 50 years cartography textbooks have had long discussions on how to achieve figure-ground through good map design,

Figure 2.1   *All Is Vanity* (1892). A woman in front of her "vanity" (make-up table) or a grinning death mask? This famous optical illusion was drawn by the American artist Charles Allan Gilbert.

but these discussions assumed that everybody would pick out the figure from the ground in the same way.[1]

Critique therefore does not seek to escape from categories altogether, but to show how they came to be, and what other possibilities there are. This sense of critique was developed by Immanuel Kant, especially in the *Critique of Pure Reason* (1781; 2nd edn. 1787). For Kant a critique is an investigation which "involves laying out and describing precisely the claims being made, and then evaluating such claims in terms of their original meanings" (Christensen 1982: 39). Kant's essay on the question of the Enlightenment (Kant 2001/1784) describes critical philosophy as one in which people constantly and restlessly strive to know and to challenge authority.

This was a radical step. At the time, most people got their knowledge from the church or from classical writers such as Plato and Aristotle. But perhaps dating from the fifteenth and sixteenth centuries and the religious struggles of the time, some people started to question these authorities. By the time of the Enlightenment therefore, there were the beginnings of a questioning attitude that would be fully developed by Kant.

This questioning attitude is not unrelated to the question of power, because it asks "what is an authority?" and "who shall have authority?" The church? The military? The government? These questions are political ones, and indicate that critique, as well as asking about the unexamined assumptions behind our practices, can also therefore open up other ways of doing things. It asks "well, we seem to be doing it this way, *but do we have to?* Isn't there an alternative?"

To return to Foucault:

> I will say that critique is the movement through which the subject gives itself the right to question truth concerning its power effects and to question power about its discourses of truth. Critique will be the art of voluntary inservitude, or reflective indocility. The essential function of critique would be that of desubjectification in the game of what one could call, in a word, the politics of truth. (Foucault 1997b: 32 translation modified by Eribon 2004)

In other words critique is a political practice of questioning and resisting ("voluntary inservitude"!) what we know in order to open up other ways of knowing. I dwell on these points here because of another misunderstanding about critical cartography and GIS which has sometimes characterized them as purely rejectionist. For example, critique is sometimes described as if it rejected all forms of knowledge or truth. The point though is not to reject, but to carefully consider the truth claims of maps and GIS (and there are a lot of such claims, as we shall see, beginning with the idea that the map is a natural reflection of the landscape). In other words, knowledge does not just exist "out there" but is created and then is privileged by being divided between truth and falsity. How truth comes to dominate is due to some fairly specific rules. Many of these rules have geographic centers, or occur at particular points in time. Critique can uncover these rules and the times and spaces in which they occur.

The modern emphasis on critique in many of the social sciences owes a substantial amount to the Frankfurt School's development of critical theory. The Frankfurt School, known formally as the Institute for Social Research, was founded in Germany in 1923 and moved to New York in 1933 when Hitler came to power. The writers most closely associated with the school included Max Horkheimer, Theodor Adorno, Walter Benjamin, Herbert Marcuse, and later Jürgen Habermas. Many of these writers sought to release the emancipatory potential of a society repressed by technology, positivism, and ideology. For example, Adorno argued that capitalism, instead of withering away as Marx had predicted, had in fact become more deeply established by co-opting the cultural realm. The mass media, by pumping out low-quality films, books, and music (and today, TV or internet) substituted for people's real needs. Instead of seeking freedom and creativity, people were satisfied with mere emotional catharsis, and were reduced to making judgments of value on monetary worth. Frankfurt School writers sought to dispel such harmful and illusory ideologies by providing an emancipatory philosophy which could challenge existing power structures.

Critical Cartography and GIS: Some Basic Principles and Examples

Critique has a number of basic principles. First, it examines the (often unexamined) grounds of our decision-making knowledges; second it situates knowledge in specific historical periods and geographic spaces (rather than being universal for all time); third it seeks to uncover the relationship between power and knowledge; and fourth it resists, challenges, and sometimes overthrows our categories of thought. The purpose of critique is not to say that our knowledge is not *true*, but that the truth of knowledge is established under conditions that have a lot to do with *power*. Critique is therefore a politics of knowledge. As Gregory put it: "critical theory is a large and fractured discursive space, by no means confined to the Frankfurt School and its legatees, but it is held in a state of common tension *by the interrogation of its own normativity*" (Gregory 1994: 10, original emphasis).

Blomley recently reviewed critical geography (Blomley 2006) and found that while it is often invoked "it is rarely nailed down" (2006: 90). It is diverse and does not depend on a single theory. For Castree (2000: 956) critical geography is used as an umbrella term for a "plethora of antiracist, disabled, feminist, green, Marxist, postmodern, post-colonial, and queer geographies." Blomley's review highlights the centrality of representation in all this: "hegemony turns, in many ways on its imaginary geographies. Denaturalizing, contesting and altering such imaginaries, then, is vital critical work" (2006: 91).

Blomley identified the following as generally characteristic of critical geography:

1. It is oppositional: it targets dominant forms of oppression or inequalities.
2. It is activist and practical: it wishes to change the world.
3. It is theoretic: it rejects positivist explanation and embraces critical social theory.

But as he remarks, there "is a remarkable confidence in the power of scholarship to reach the benighted, and in the transformative capacities of people to overcome alienation through reflexive self-education" (Blomley 2006: 92), and it should be remembered that academic scholarship only takes you so far. Critical cartography and GIS is only in part a matter of scholarship, for the other half of our one–two punch, we have to turn to actual interventions, protests, transformations, and community mappings. Included in this would also be art work, blogs, mashups, and the "geospatial" web.

What has been going on in cartography and GIS that makes it critical? Geographers who have not been paying close attention and think of cartography as a technical field that has produced one or two interesting critical articles in the last 20 years are woefully out of touch. In fact cartography is a rich transdisciplinary field with something of a history of critique. This tradition of questioning has undergone an amplification in the last two decades and has perhaps finally achieved some disciplinary purchase. But this tradition has always appeared at the margins, outside the main textbooks, and sometimes outside academia altogether.

Given what we have seen so far it is possible to distill four principles or strands of critical thought in cartography and GIS. The purpose here is to sort through the rich variety of work, but there's also a danger that these strands are taken to be definitive rather than the flexible suggestions they are meant to be. The emphasis on these strands varies from one work to another, but they can usually be found to some degree. I will be coming back to these strands time and again throughout this book. You can compare these four principles of critical cartography to the more general ones that Blomley identified for critical geography.

1.   The first principle of critical mapping is that maps are incredibly useful ways of organizing and producing knowledge about the world, but that these orders of knowledge also incorporate *unexamined assumptions* which act as limits which deserve to be challenged.

2.   One way to challenge these orders of knowledge is by putting them into historical perspective. This *historicization* of knowledge not only shows that other times did things differently, but by providing an intellectual history it allows us to see the edges of our own limits, and to conceive of other knowledges that might be useful. Critical mapping also emphasizes that the way maps and spatial knowledges have been deployed has varied tremendously between cultures and places. This can be described as a *spatialization* of knowledge.

3.   Critical mapping also holds that geographic knowledge is shaped by a whole array of social, economic, and historical forces, so that knowledge does not exist except in relation to *power*. When we speak of maps as political, it is this relationship between knowledge and power that is at stake.

4.   The critical mapping project is also one which has an activist, *emancipatory* flavor to it. Sometimes this approach is concerned with overthrowing the influence of official knowledges (such as those of the government or the state) by showing their historical and spatial contingency (Livingstone 2003; Sparke 1998). At other times this approach seeks to dismantle more specific forms of knowledge, such as recent work by feminists in critical GIS or community activism in participatory GIS (Elwood 2006b; Kwan 2002a; Schuurman 2002).

These four principles are meant to act as a guide to a deeper appreciation for the critical project in mapping and GIS, rather than a definitive categorization. They overlap not only with each other but as we have seen with other critical work. Rather than going through each principle in turn I will aggregate them slightly and discuss how they are used in two larger fields, that is, in theory and in practice. This division is not meant to mark a definitive division between theory and practice – we don't wake up and decide to do theory one day and practice the next. Rather they are aspects of the critical project which are part of each other.

Theoretical critiques

Critical cartography assumes that maps *make* reality as much as they represent it. Perhaps John Pickles expresses this best when he says:

> instead of focusing on how we can map the subject . . . [we could] focus on the ways in which mapping and the cartographic gaze have coded subjects and produced identities. (Pickles 2004: 12)

Pickles rethinks mapping as the production of space, geography, place, and territory as well as the political identities people have who inhabit and make up these spaces (Pickles 1991; 1995). Maps are active; they actively construct knowledge, they exercise power and they can be a powerful means of promoting social change.

Increasing attention was paid to how maps inscribe power and support the dominant political structures. Wood's *The Power of Maps* (1992) was particularly significant in this regard. It was both a major institutional exhibition at the Smithsonian and a bestselling book (e.g., it was a Book of the Month selection). It exerted a considerable influence on academics and non-academics. Wood argued that maps express the interests of certain groups and that these interests are not always explicit. But Wood was no conspiracy theorist, he showed that the map interests could be made to work for others. This was a very well received argument and it proved something of a manifesto for many counter-mapping projects (see Figure 2.2).

The Smithsonian exhibition included many exhibits that would not normally be thought of as "political" or "interested" maps. One of the most popular was Wood's analysis of the North Carolina road map. This is the kind of map given away for free at rest stops at various places on American interstates. It is not the kind of map you would suspect of harboring hegemonic purposes! In fact, Wood showed that through a combination of inclusions (such as the state governor and his family standing next to a large car; while the back of the map was covered in adverts for local businesses) and exclusions (any depiction on the state map of paths, bike lanes, or public transport) an image of the state as pro-business and car friendly was created. The resulting impression was that North Carolina was a good state in which to live or invest (part of the importance of this is that the state gives these maps out at rest-stops *just inside* the Carolina border). Wood's analysis here was very much influenced by those that were performed by the French author Roland Barthes (e.g., Barthes 1972) in which he took everyday objects (the *guide bleu* travel book for example) and revealed their hidden meanings.

Figure 2.2   Power of Maps Smithsonian catalog, brochure and children's activity kit.

Wood's reappropriation of the map's agenda was a significant move for a least two reasons. On the one hand, it showed that maps did not just have to serve the state, although they obviously did so in the past and continue to do so (Buisseret 1992). Maps could also work for "the people," a theme that has been at the heart of not only the recent surge in map hacking and the geospatial web but of participatory GIS as well. Mapping became something that could actually be used to resist the state, especially in its guise as an authoritative power. The weapon of the map could be turned to other ends beyond those of the state. It is significant that this understanding of mapping (that it exists not just for the state) is also an understanding of power. Whereas previously we might see an opposition between the state as the locus of power on one side, and ordinary individuals in the other, what Wood alludes to is that power may circulate from below just as much as from above.

Turnbull (1993), for example, includes the story of a map of Aborigine Dreaming trackways in the Great Victoria Desert. In 1981, Kingsley Palmer of the Australian Institute of Aboriginal Studies in Canberra collected information about the myths and trackways in this part of Australia. Palmer then transferred this information to a Western topographical map and took the map with him as a gift to the Pitjantjatjara Aborigine community, who responded to the map with great interest. In fact, the community regarded the map as extremely precious, and full of secrets that should be known only to them (a number of the myths Palmer had inserted on the map should only be discussed by grown men of the community). So it was agreed to put the map in a bank vault in a nearby town, where it could only be withdrawn with the community's permission.

One of the reasons for this was that the community was involved in a long land dispute with the government to decide if the lands would be returned to the community. In order to resolve this question, the government flew some representatives to meet with members of the community. Palmer was also invited, and when he got there he says he was found that his map had been withdrawn from the bank vault, and "at a suitable time when the men had taken the parliamentarians to one side, the map was unrolled on the desert sands . . . the Aboriginal people were at great pains to point out the extent of the Dreaming tracks and the numerous sacred sites that were noted on the map" (Turnbull 1993: 60).

As a result, the lands were returned to the community in 1984 and the map was kept as a kind of title deed in the local bank. Thus although it was drawn by an outsider to the community, indigenous people were able to successfully use it in a struggle against the state in a land ownership case.

Both these examples rely on critical theory to deal with relations of power, but notice also that they have practical ends. Wood points out that if maps are powerful, that power can be used by anybody, not just those in powerful positions. Turnbull's account of the Pitjantjatjara Aborigine community is an example of how maps can be used against the state.

One scholar's work has been of undeniable importance in furthering the theoretical development of critical cartography. In a series of papers toward the end of his life, Brian Harley brought into the discipline the ideas of power, ideology, and surveillance, arguing that no understanding of mapping was complete without them (see Chapter 7). Harley brought these ideas into cartography from its margins, and often indeed from well outside it. Edney has pointed out that Harley was well read in radical human geography (Edney 2005a), and Harley also scattered his texts with references to Derrida, Roland Barthes, and Foucault. Harley therefore represents an opening of the discipline of cartography to extra-disciplinary ideas in a way that had not occurred since the immediate post-war years when Arthur Robinson imported ideas from behavioral psychology and architecture.

Rejecting the binary oppositions until then dominant in cartography, such as art/science, objective/subjective, and scientific/ideological, Harley sought to situate maps as social documents that needed to be understood in their historical contexts. Harley then argued that mapmakers were ethically responsible for the effects of these maps (Harley 1990a). In this way he could explain the dominance of seemingly neutral scientific mapping as in fact a highly partisan intervention, often for state interests.

Other writers took up this last point and applied it to the field of GIS. GIS practitioners responded in kind, accusing social theorists of ignoring the tremendous insights possible with GIS (S. Openshaw 1991), and of attacking one of the few real contributions of geography beyond the discipline. For a few years the arguments constituted geography's own version of the "culture wars." However, as Schuurman has documented, there was a strong vested interest in reconciliation, which has resulted in some recognition of the validity of each other's arguments (Schuurman 1999b; 2000; 2004). During the 1990s there was an effort to develop a more social or critical

GIS. The most notable of these is that GIS has been taken outside the academy and used for community participation (Craig et al. 2002). As yet however, there has been little uptake of social GIS from human geographers despite the fact that GIS plays a large role in social decision-making such as public health analysis (Schuurman and Kwan 2004).

So the theoretical critiques of the 1980s and 1990s did not arise from nowhere. They were made possible and given strength by the fact that throughout its history mapmaking has butted up against marginalized and local knowledges. While the history of cartography during the twentieth century is one of increasing scientific aspirations, there has all along been a parallel series of mappings that were not scientific. As the ongoing *History of Cartography* project has repeatedly shown, indigenous, pre-scientific, or simply non-disciplinary mappings (that is, those developed outside the confines of the cartographic discipline) abound in many human cultures. In Volume I of that series founding editors Harley and Woodward adopted a more expansive definition of the map in order to include examples of maps that did not fit with textbook cartography: "maps are graphic representations that facilitate a spatial understanding of things, concepts, conditions, processes, or events in the human world" (Harley and Woodward 1987: xvi). Such a definition places emphasis on the role of maps in human experience, rather than the look or form of maps (as had previously been typical [Robinson 1952]). Harley and Woodward opened the door to many non-traditional and non-Western mapping traditions. Their project, with its consideration of hundreds of new examples of maps, almost certainly informed Harley's theoretical work, and not the other way around (Edney 2005b; Woodward 1992a; 2001).

Critical mapping practices

If the theoretical critique cleared conceptual space for alternative mappings it has fallen to a variety of practitioners, most of them from outside the academy, to explore what this meant in practice. Here we need to focus almost entirely outside of academic cartography and GIS. Two developments are especially notable: artistic appropriations of mapping and the storming success of map hacking, mashups, and the geospatial web. Each of these will be discussed in more detail later (Chapter 3).

The artistic community has long experimented with maps, their meaning as representations and as efforts to find our place in the world (Casey 2002; kanarinka 2006). The philosopher Edward Casey argues that in the last 50 years mapping and art have experienced a dramatic convergence:

> On the one hand, ways of painting have developed that can be considered mapping – not just incidentally or partially, but through and through. On the other hand, a new art form has evolved, that is, earth works, which map by their very essence and not just exceptionally. (Casey 2005: xxii)

Wood has pushed the dates back even further, citing nearly a hundred years of map art (Wood 2008). Many of these artists are interested in geographical re-mappings, and have worked with the assumption that maps are political without explicitly saying so. This artistic appropriation of the politics of representation has long historical roots, from the avant-garde artistic movements at the turn of the century (Georges Braque, Paul Cézanne) to the Situationists and "psychogeographers" of the 1950s and 1960s. These latter groups sought to radically transform urban space by subverting cartography as part of a project of political resistance (Pearce 2006). Their "subversive cartographies," by assuming that cartography was always already political, created different arrangements of space (such as the famous 1929 surrealist map of the world [Pinder 1996; 2005]. See Figure 2.3).

In this map, the USA is omitted except for a hyper-sized Alaska that faces off against Russia (shades of Sarah Palin!). The territory where it "should" be is occupied by "Labrador." Greenland and Russia appear in exaggerated sizes, reminiscent of their distortions on many maps. The equator is labeled, but seems to meander at will through a maze of islands in the Pacific Ocean. South America is shrunken and consists only of Peru and Tierra del Fuego, Mexico replaces the continental United States (a reference to Frida Kahlo?). Each coastal outline is largely devoid of any interior (only two cities are named; Paris and Constantinople: someone took a trip on the Orient Express?). The line work is wavy and uncertain, as if the artist was elderly or bored. The map is centered not on a country, but on empty ocean.

As with the Frankfurt School, part of the Situationist critique was that modern society's basis in consumer capitalism caused deep alienation. Guy Debord's book *The Society of the Spectacle* acts as something of a guide by emphasizing that everything has become represented and thus devalued, everything is a media spectacle (Harmon 2004). This work has produced a tremendous legacy, aided by the infusion of mapping technology in the late 1980s which set the stage for an explosion in "locative art" and psychogeographical mapping (Debord 1967/1994). Lee Walton averaged all the coordinates on a tourist map of San Francisco to come up with a single "Average Point of Interest" where he installed a bronze plaque (Casey 2002; Denis Cosgrove 1999; 2005; Harmon 2004). These "map events" challenge the commensurability of Euclidean space, a basic assumption of much GIS. That is, if you import a map of Copenhagen into your GIS it will georeference itself into a European space, and will not be overlaid on New York City – the two spaces are physically separated by the Euclidean coordinate system. If you break from Cartesian space what new perspectives are thrown up? What strange conjunctions and serendipitous new knowledges? Like the surrealist map the answer to these questions is not a distorted map, but an impossible one, yet one that exists and can be created. Perhaps it is better to say it is a paradoxical map.

These few examples could be multiplied, but the bottom line for the moment is that the disciplinary field of knowledge, cartography, that has corralled maps and mapping practices for half a century is undergoing a transition. Some see this as a result of the rise of GIS and the spatial database (GIS "killing" cartography to put it in stark terms). Others see it as the result of the closure of geography departments

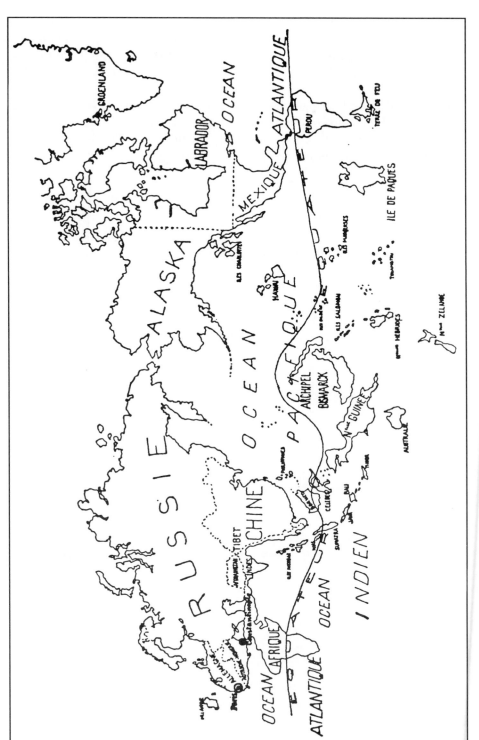

Figure 2.3 Surrealist map of the world [Paul Eluard?] 1929. Source: Waldberg (1997).

in Europe and the Americas. But there is a larger picture too, and that is that mapping has passed beyond the hands of the old discipline. It's passed beyond it in terms of critical theory and in terms of mapping practices. So while the cartography discipline may be in disarray, mapping has never been healthier. It is this paradox that confronts us today.

For GIS and critical cartography, founded in a post-war sensibility of internal empiricism and the map communication model, the social relevance critique has proved a difficult one to absorb. In subsequent chapters, we shall delve more deeply into where and how this legacy arose.

Note

1   The first mention of figure-ground in a cartography textbook is in Robinson (1953) who appeared to derive it from work in psychology.

Chapter 3

Maps 2.0: Map Mashups and New Spatial Media

"In the process, [map amateurs] are reshaping the world of mapmaking and collectively creating a new kind of atlas that is likely to be both richer and messier than any other."
                                    The New York Times (Helft 2007)

"What's happening now . . . is that instead of just GIS experts talking to each other, or experts making maps for regular people; regular people are talking to each other, and they're making maps for each other. And that's very important. . . . the story of the where is very important."
                        Michael Jones, CTO of Google Earth at GeoWeb 2007

Introduction: The Story of the Where

As the two quotes above indicate, there is an explosion of interest in mapping tools for regular people and map amateurs. These "peasants" as they have been described (Perlmutter 2008) are a long way from the traditional image of the expert map-maker with his qualifications and specialized tools. They are ordinary people from all walks of life who nevertheless want to share their lives with friends. And they know for that, they'll need to include the geographies of their everyday lives. As Michael Jones puts it: "the story of the where is very important."

In the previous chapter we examined how critical mapping has provided a one–two punch to traditional notions of mapping. One of these blows was landed by a relative upstart on the mapping scene. In this chapter we will take a closer look at the phenomenon of map mashups and the strangely named concept of the "geospatial web."

The crux of the challenge that these recent developments pose lies in the way that cartography has long been practiced. For most of its history mapping has been the practice of powerful elites – the so-called "sovereign map" (Jacob 2006). The sovereign map for Jacob refers to the fact that the map was a dominant political force; one which held sway as a way of knowing the world (even as we look back on them now and see just how inaccurate they were). Maps held power; they *were* sovereign.

But maps were sovereign also in the sense that it was literally only sovereigns and those in power who made and used maps. Maps were elitist and were made for elites. Nation states, governments, the wealthy, and the powerful all dominated the production of maps (Buisseret 1992). For example Buisseret tells the story of a Lucayan Indian who was brought back to Spain by Christopher Columbus and presented to King Ferdinand. The Indian was able to lay out on the table a rough map of the Caribbean using stones as markers, and it seems likely that Columbus and his crew benefited from this indigenous knowledge in making their maps for the royal court (Buisseret 2003).

Map sovereignty is now being challenged by the emergence of a new populist cartography in which the public is gaining (some) access to the means of production of maps. This is certainly not an isolated development. It is part of a larger movement of counter-knowledges that are occurring in the face of ever-increasing corporatization of information such as the consolidation of the news media into the hands of a few global multinationals and their dominance by fairly narrow interests. The internet and web, blogs, and the netroots (online political activism) are all reasons for this "people-powered" control of information (Armstrong and Zúniga 2006). In this chapter I focus on some of the exciting new developments that can help create, visualize, and disseminate geographical information. And yet at the same time many of these developments are in the hands of quasi-monopolistic media companies (Google, Yahoo, Microsoft). We'll ask what it means for traditional expert-driven GIS and mapping if "map amateurs are reshaping the world of mapmaking." Is there a tension between this and traditional GIS? What do these tools offer – only visualization and sharing or analysis as well? And if so, will they transform or even replace GIS as we know it? In sum: is there a new politics of knowledge?

Because the field is so new, there is not yet a single, catchy name for these online mapping tools and services. Some of the suggestions include map mashups, map hacking (Erle et al. 2005), the geospatial web or geoweb (Scharl and Tochtermann 2007), neo-geography (Turner 2006), locative media, volunteered geographic information (Goodchild 2007), DigiPlace (Zook and Graham 2007b), and new spatial media. I would not like to say which (if any) of these will eventually be adopted. I've elected to mostly call it the geoweb or new spatial media.

The Google Experience and the First Mashup

Claudius Ptolemy, Gerardus Mercator, Christopher Columbus, Lewis and Clark . . . and Paul Rademacher.

Who?

Paul Rademacher may be the most influential cartographer of the twenty-first century, but you've probably never heard of him. He's not famous. Still, his achievement ranks alongside those of the great cartographers and explorers. Paul Rademacher invented the first successful map mashup.

Rademacher is not a cartographer. He was an animator for DreamWorks, the film studio that makes the *Shrek* movies. In late 2004, Rademacher was looking for an apartment in the San Francisco Bay Area using Craig's List, the classified housing listings that was founded in 1995. As he drove around with piles of printouts and maps, Rademacher thought "wouldn't it be better to have one map with all the listings on it?" (Ratliff 2007: 157). His timing was excellent. On February 8, 2005, Google Maps went online and within only a matter of hours programmers had reverse-engineered it so that their own content, *rather than Google's*, would appear on the maps (Roush 2005). What this meant was that Google Maps had been hacked – not by mischievous trouble-makers, but by people who wanted to use Google's well-designed maps to display and share their own data.

Map hacking is the practice of exploiting open-source mapping applications or combining one site's functionality with another's. These are known as "mashups." A mashup is a website or web-based program that combines two or more sources of content into one tailor-made experience (Butler 2006; Miller 2006; Wikipedia 2007). These exploitations are possible because of something called extensible markup language (XML) and application programming interfaces (APIs). Open-source APIs define the way one piece of software connects up with another. Think of them as public interfaces. Many online mapping applications have APIs, including Yahoo Maps and Google. Google's map interface allows users to integrate data using only ten lines of code (Butler 2006). The Google API allows other data to be fed to it and displayed as a Google map.

In June of 2005 Google released Google Earth (GE), which uses the same underlying dataset (essentially a very detailed set of imagery of the entire earth), but which projects it onto a gorgeous-looking interactive digital earth. Both Google Maps and Google Earth were huge hits with the online community. Google claims GE has had more than 400 million unique activations, and the Google Earth community, a place where members share interesting geographic information, has over one million registered users.

Seeing all this, Paul Rademacher remembered his house-hunting experiences the previous fall, and he developed what would become recognized as the first successful map mashup. Called HousingMaps it combined Google Maps with Craig's List to provide the kind of online experience he had needed a few months earlier:

> One Thursday night, he posted a link to the demo on craigslist, and by the next day thousands of people had already taken it for a spin. "I had no idea how big it would be," he says. (Ratliff 2007: 157)

The site, like most mashups, is a hybrid of different data that is sent to Google Maps for display. The word "mashup" is a borrowed term from the music industry, where it refers to the combination of two or more different songs to create a third, new song. With the advent of computers many groups now produce remixes or alternative versions of their original song, but the mashup goes one step further by combining two completely different sources. With the advent of the reverse-engineered Google maps, it was now possible to deliver just about any spatial information you could conceive of via their maps. Map mashups were a significant advancement in people-powered mapping.

Google's clever insight into this development was not to try and prevent it, but to endorse it by deliberately opening up the part of the code that allows people to hang their mashups onto the maps. In June 2005 Google released the API so that reverse engineering would no longer be necessary. By putting put their weight behind "open-source" software for maps, Google's return is that they are now the second-most trafficked mapping website (MapQuest is number one) even though their business is not solely geospatial.

In the fall of 2005, Hurricane Katrina hit New Orleans, and millions of people used Google Earth to visualize the area and obtain information that was not otherwise available (Ratliff 2007). The US government agency NOAA posted hundreds of freshly flown aerial imagery on GE, and the exploding user community posted photographs and personal accounts of the floods; GE also provided rapid response imagery updates. Katrina was a big test for the nascent Google Earth. By flocking to GE to learn more about it, and by using it to help each other out, its profile was raised at a time when Americans were looking for answers about Katrina. No less remarkable was that the American government used GE as a distribution tool, rather than their own usual channels – they essentially admitted that GE was better.

More recently, Google have released MyMaps, a kind of do-it-yourself map mashup capability (April 2007) and "mapplets" or map applets (May 2007). MyMaps allow anybody to annotate a Google Map with a location, route, or area (polygon) and to save and share the map (they can also be turned into Google Earth files known as KML files). Mapplets are special pieces of code that allow extremely varied applications to be added into Google Maps – essentially they are like little GISs. As for Paul Rademacher, his story had a happy ending: Google hired him.

Free and Open Source Software (FOSS)

While these developments benefit from Google's API, a lot of them have come out of the free or open software movement. Google itself is not making any mashups. The basic idea of open source software is that it is "configured fundamentally around the right to distribute, not the right to exclude" (Weber 2004: 16). Open source proponents talk about free in the sense of freedom, or *libre* not *gratis*. Richard Stallman, an early and ardent advocate of free software in the "libre" sense, describes four essential freedoms:

- the freedom to run the program for any purpose;
- the freedom to modify the program at the source code level;
- the freedom to distribute copies (sell or donate);
- the freedom to distribute modified copies (Stallman 1999: 56).

These freedoms are deliberately contrary to international copyright law which protects both the original work and any derivatives made from it. (The exception is the US Federal Government whose works are not copyrighted, but this exception does not apply to data created by others countries – the Ordnance Survey in the UK for example.) In the map world this has meant that if you own the copyright on your map other people cannot use any part of it – even if it is combined with other new data – without your permission. Since in practice most maps are made from other maps, many people feel this has served to stifle innovation. Probably the strongest feelings about this have been aimed at the UK's Ordnance Survey, a governmental body that has been charged with full cost recovery.

The Ordnance Survey's closed source effect on the availability of geospatial projects is best seen in the collapse of the Virtual London Project which was developed by the Centre for Advanced Spatial Analysis (CASA) at the University College London. The CASA project was designed to place over three million of London's buildings in three dimensions online via Google Earth. Technically, the project succeeded. Working over a period of about six years, CASA produced 3D models of the buildings in outline and with photographically realistic facades of an area about 2,000 square kilometers in size at 1-meter resolution. This model can be populated with almost any kind of data (live pollution readings, for example). Because the model was created in part from Ordnance Survey data, a license was required to make this available via Google Earth. Although Google was willing to pay the OS for this license, they preferred a fixed-rate fee, rather than the pay-per-use fee demanded by the OS (Cross 2007). In August 2007, CASA announced that the project had been discontinued. (Other city 3D projects are being developed in Dresden, Berlin, and Hamburg.)

Stallman went on to found the Free Software Foundation which instituted a special kind of software license known as the General Public License (GPL), as well as concepts such as "copyleft" and Creative Commons licenses. Because the word "free" can be misleading however, the more common name for this software is open source or free and open source (FOSS). In 1997 the open source concept reached a wider audience in a well-distributed work "The Cathedral and Bazaar," which later appeared as a book (Raymond 2001). Histories and the appraisals of open source are numerous (DiBona et al. 2006).

A key insight possessed by the FOSS movement is that it can more easily create and share information than a closed source, such as traditional GIS. While these capabilities are based on geospatial technology, the point is that they did not spring from the disciplines of cartography or GIS. They have been developed by programmers intrigued by mapping's potential to deliver meaningful information. Indeed, it is rare to find references to cartographic literature at all in these new developments

(Bar-Zeev 2008; Miller 2006; Zook and Graham 2007a). Yet providing accessible and inexpensive mapping tools may radically reshape how mapping is done (Fairhurst 2005; MacEachren 1998; Taylor 2005).

The bottom line is that map mashups have become trivially easy to make and, more important, much more visible. They can be shared and embedded in other webpages as "live" map services (i.e., not just as images of maps, but with the ability to zoom, pan, and query) through the use of keyhole markup language (KML). KML is a file for sharing geospatial data, along with GeoRSS, both of them based on a common standard web format known as XML (extensible markup language). Many of these standards are coordinated through the Open Source Geospatial Foundation (OSGEO).

If the acronyms bespeak a strong technical element of these developments, another effort epitomizes the fuller societal implications of open source mapping. The Open-StreetMap (OSM) is nothing less than a project to independently map the world. It started in the UK because of closed source mapping providers like OS. Participants use GPS units to capture waypoints along roads and streets, as well as around area features like parks. They then upload these, symbolize and label them, and add them to OSM dynamic "slippy map" style database. These data are subject to the Creative Commons license, making them open source and therefore usable by others for free (with proper attribution). While this may sound like a daunting task, progress has been remarkable and several areas of the world (mostly in Europe) have been "completed."[1]

The OSM benefits from another benefit of the crowd: the fact that a little contributed by many adds up to a lot. OSM draws its data from a myriad of people hiking, biking, driving, or taking the train who carry GPSs with them. No person, no team could provide enough data by themselves. OSM is the Wikipedia of mapping.

Political Applications

"*Republicans still control the maps.*" Chris Bowers, MyDD.com, October 2006[2]

There is now some intriguing evidence that suggests that access to, control of, and dissemination of geospatial information is changing political participation (Talen 2000). While much political discussion occurs in the traditional or "mainstream media" (MSM), much is now also held in the emerging arena of blogs. Blogs now constitute a significant and noteworthy component in today's political landscape (Perlmutter 2008). Blogs and online political activism (the netroots) play important roles in getting out the vote (GOTV) and online fundraising. Since the 2004 elections and the success of Howard Dean and organizations such as MoveOn.org, the intersection of the so-called "netroots" (Armstrong and Zúniga 2006: 2) and politics has only become stronger.

Alongside the netroots (a pun on grassroots) and often in conjunction with them are a range of mapping and GIS tools now available for the public. These tools often rely on making linkages between different kinds of knowledges; for example between different sources of data (such as Google Maps and the Federal Election Commission), and different software programs (such as between GIS and Google Earth). These linkages, effected through open source software and APIs, mark a potentially new phase of political activism and collaboration characterized by more democratic access, control and production of information and knowledge, a more local "micro-politics," and potentially a way to break the hold of establishment "big money" incumbents.

A sense of the influence of political blogging can be gained by considering the following achievements:

- **Investigation**: Talking Points Memo (TPM), a New York City-based blog, won a George Polk Award in February 2008. The Polks are often described as the Golden Globes of journalism, and TPM's award was the first to a blogger. During 2007, TPM aggressively pursued and broke news concerning the United States attorney firings, which turned out to be politically motivated by the Bush administration. As a result a senior official in the Department of Justice resigned, and the Attorney General himself, Alberto Gonzales, stepped down under a cloud.
- **Participation**: The proportion of Americans using the Internet as their source of political information has sprung from less than 10 percent in 2002 to about 60 percent in the 2008 Presidential election (Rainie and Horrigan 2007; Smith 2009; Smith and Rainie 2008). While television still dominates, the Internet has now eclipsed newspapers.
- **Fund-Raising**: This has also burgeoned massively. Between its launch in 2004 and spring 2009 for example, a single online website, ActBlue.com, raised over $100 million for 3,200 Democratic campaigns from 420,000 contributors. Republican candidate Ron Paul raised $4 million dollars or more online in a single day on more than one occasion during his presidential campaign, mostly through small contributions.
- **Popularity**: Political blogs regularly rank among the most-visited blogs on the web of any kind. DailyKos gets at least three-quarters of a million visits a day and often more than a million. Perhaps more impressively, because it is a community rather than the work of an individual, it has produced nearly half a million diary entries, and over 15 million comments since 2003.[3] The Huffington Post (#4), DailyKos (#11), Think Progress (#26), Crooks and Liars (#33), Drudge Report (#39), Talking Points Memo (#42), and The Daily Dish (#47) are political blogs in the Technorati 50 Most Authoritative rankings (i.e., most linked to) on the Web.
- **Organization**: Social network sites such as Facebook have a strong utility for political organization. Successful organization often depends on communicating effectively with your members, for example in a labor dispute with management, in getting out the vote, or in organizing caucusing or campaigns. Facebook and MySpace groups can sign up large numbers of people, send out mass emails, and provide key facts and information. With American union membership holding steady or rising slightly in 2008 (to 12.6%) online organizing has much potential. The Trades Union Congress, National Union of Journalists, and other unions have all used Facebook to organize, though as yet the results are mixed.

- **Culture jamming**: Activists interested in pursuing resistance to commercialism have several web-based tools at their disposal. Adbusting can be achieved by buying Google ads for a company, so that anyone searching for it will see anti-corporate ads on the results pages. So-called "cyber pickets" were used during the Writers Guild of America (WGA) strike in the USA in 2007–8 to plaster pro-WGA graphics all over the shows on the major networks.

And what of political geography? For example, the FairData website provides community-based interactive maps down to the precinct and census block group level for the whole nation.[4] These data are linked to open source mapping APIs such as Google Maps for visual display. Users can pan and zoom across the maps and display different layers of information (the site uses a sophisticated online GIS as a backend to the web pages). For a GOTV effort, community organizers can create maps of the number of non-voters by precinct. In this map of Philadelphia, the voting turnout is shown for each precinct, allowing the GOTV team to target non-voting neighborhoods (see Figure 3.1).

The map shows that turnout varied quite considerably across the city, and in many areas was below 40 percent. These can then be targeted by the GOTV effort. The maps can also show individual households that did not vote for even more targeted efforts. As far as I am aware these are the first tools available to the public that were previously only compiled in secret party political precinct maps.

Other geographical tools have also been devised to get out the vote or to manage organizations. These include Catalist, which uses a web-based map front-end for accessing and integrating data (e.g., from the Census) and for producing voter contacts for field workers, and the Donkey. The latter is:

a volunteer management program developed in 2005, along with bar-code scanners, Palm Pilots and Google Maps, whose satellite feature allowed field organizers to cut turf without having to physically explore the routes, produced huge efficiency gains. (Stoller 2008: 22)

Other efforts that have proved effective are to better generate voter files following the Howard Dean Democratic campaign in 2004. One of the issues there was that there was no shortage of volunteers, but their help couldn't be easily assimilated. As Stoller remarks "if it takes all night to prepare a map for a canvasser, you can't absorb very many volunteers, and you can't talk to many volunteers face to face" (Stoller 2008: 22). (Studies consistently show that face to face meetings are more effective than mailers or phone calls, and a good field or ground campaign can swing voting by as much as 3 to 5 percent.) For Stoller, politics is about networks:

Political power is more and more situated in far-flung networks that can be activated and deactivated quickly, and the new millennial generation that will be the political backbone for the new progressive America likes it this way. (Stoller 2007: n.p.)

Figure 3.1 Turnout and race in Philadelphia. Source: Bill Cooper, www.fairdata2000.com. Used with permission.

For people like Stoller, the political landscape is shifting. Where television was the means of persuasion in the past, this is fading and becoming less important, due in part to the atomizing of TV viewership:

> Social networks like Facebook, Blackplanet, blogs and SMS, as well as basic e-mail, can be layered onto the clean new databases to reach voters wherever they are, for much less than TV advertising. (Stoller 2008: 23)

Do these tools by themselves mean that the political landscape is now more democratic? Not necessarily. As Foucault observed, power and knowledge go together, and nowhere is this more salient than the relationship between digital mapping and geovisualization with the military. The size of the military investment in GIS, such as the geospatial intelligence community (GEOINT), is not known, but was formally recognized in the creation of the federal National Geospatial-Intelligence Agency (NGA) in 2004, and the military's doctrine on GEOINT has been described in recent reports (United States Joint Forces Command 2007). Because GIS has historically been largely associated with government and industry (e.g., the 2006 GEOINT 2006 Symposium was keynoted by the Director of National Intelligence, John Negroponte), there are many who view GIS as being just another mechanism of government control and surveillance (Pickles 2004; Smith 1992). Pickles argues that many of the new mapping capabilities are wonderful:

> They provide more powerful tools for local planning agencies, exciting possibilities for data coordination, access and exchange, and permit more efficient allocation of resources, and a more open rational decision-making process. (Pickles 2004: 148)

Yet these systems are taking place in a larger context of economic production and a "culture of military and security practices" (Pickles 2004: 152). Trevor Paglen, a geographer at Berkeley, has investigated many of these "hidden geographies" and even provided a map mashup of the CIA's "extraordinary rendition" flights (Paglen 2007; Paglen and Thompson 2006).

Professionals Versus the Amateurs: De-professionalization or Re-professionalization?

There has, of course, been if not a backlash, then a hesitation or fuller reconsideration of the effects of the geoweb.

If crowd-sourced mapping is like Wikipedia, that comparison worries a number of people. Wikipedia after all is formed from the contributions of tens of thousands of people, none of whom have been vetted or asked to show any credentials (or even identify themselves). If as the cartoon has it "on the Internet no one knows you're a dog" then on Wikipedia no one knows if you're an expert.

From Wikipedia's point of view that's precisely the point. Wikipedians accept that while not all entries are correct there is an inherent self-correcting process performed by people with a vested interest in the content of the article. And it is an encyclopedia, not an academic journal and thus reflects our culture, warts and all (the entry on Pamela Anderson is longer than the one for Hannah Arendt according to one critic). But parts of it do deal with scientific matters, so the question arises how we assess this material given that students are increasingly turning to it?

In 2005 the British journal *Nature* decided to assess the accuracy of Wikipedia for itself. It asked a group of area experts to assess content within their expertise from both Wikipedia and from the online portion of the *Encyclopedia Britannica* (Giles 2005). The experts determined errors in articles without knowing from which source they came. Remarkably, reviewers found only eight serious errors, four from each source. They also found that the overall error rates (including factual errors, misleading statements, and omissions) were similar: four in each Wikipedia article and three in each *EB* article.

Nature's study goes some way to forming a picture of the quality of new media content, though it is surely far from being the last word. No similar studies have been performed on geospatial data, for example comparing OSM to Ordnance Survey maps.

The study is also unlikely to allay fears of those who feel that open source mapping is de-professionalizing the geospatial field. Critics of "amateurism" point out that because there are no controls on quality the result is the Internet is awash in low-quality content – not just in the geospatial industry but as a whole. The journalist Andrew Keen for example recently issued a strong condemnation of the "cult of the amateur," identifying it as being responsible for the proliferation of blogs, YouTube videos, the fragmentation of identity, and the betrayal of ethics (Keen 2007).

Another criticism is that proponents of the geoweb may be shooting themselves in the foot by emphasizing its amateur status. If, after all, anyone can deploy these ready-mades, then what future for the profession not just of cartography/GIS but geography? The expertise presumably held by geographers is no longer needed, or at least no longer recognized. What obtains to this situation then, is a sort of critique of expertise, its necessity, and relevance. From the experts' point of view, this is a threatening development.

Online maps wiping out history?

This criticism got a more direct airing in a widely reported talk in the fall of 2008 given by Mary Spence, the President of the British Cartographic Society. According to Spence, online maps are destroying Britain's heritage:

> Corporate cartographers are demolishing thousands of years of history – not to mention Britain's remarkable geography – at a stroke by not including them on maps which millions of us now use every day. We're in real danger of losing what makes maps so unique, giving us a feel for a place even if we've never been there. (BBC 2008)

The implication of this, she felt, was that future generations of map readers would suffer from inferior mapping. Already we can read headlines like "Fifty Per Cent of Drivers Cannot Read a Map" (Massey 2007).

Spence added "But it's not just Google – it's Nokia, Microsoft, maps on satellite navigation tools. It's diluting the quality of the graphic image that we call a map" (BBC 2008). Ironically although this talk created quite a stir in the geoweb, Spence also pointed to efforts such as the OpenStreetMap as a response to these cartographic erasures.

Possibly as a response to these and other issues, there is a growing movement in the US toward certification and approved "bodies of GIS knowledge" (DiBiase et al. 2006). The GIS Certification Institute (GISCI) is a national body which accredits "GIS Professionals" (GISPs). The governing body of GISCI includes representatives from the AAG and from its parent organization URISA. Some 1,500 people have been accredited, according to the most recent data on its website.

If the GIS wars during the 1990s were about the kinds of knowledge or epistemologies produced by GIS (e.g., positivist) then those wars in the 2000s are about ontology: should mapping be closed or open source?

A further question lies in the way users consume geospatial data. Will they use the geoweb with discernment and with critical evaluation? How can users acquire the skills to do that (are there educational implications?).

Jack Dangermond, the CEO of ESRI, reflects the uncertainty and doubt about the effects of the geoweb on professionals. From his perspective user-generated content is dubious: "He worries that even the best-intentioned amateur could provide inaccurate data that could lead to a disaster. 'Who wants to dig a hole and run into a pipe?' Dangermond asks" (Hall 2007). The conference attracted the attention of *Computerworld*, which wrote:

> The debate about whether GIS is a domain for experts or the rest of us raged throughout last month's GeoWeb 2007 conference in Vancouver, British Columbia. According to Michael Jones, Google Earth's chief technologist, by giving everyone access to GIS tools, you'll end up with "a big number of users converging on a truth." Locals, he insists, are closer to most GIS data than experts and have a vested interest in its accuracy. (Hall 2007)

It seems then that the battle lines are being drawn. Big GIS claims it is more than visualization and "eye candy" because it can do modeling and analysis. FOSS cartographers claim they are doing more and more of that too, as well as providing a true 3D world (with 3D buildings for example that can be used in urban planning). Nor is GIS particularly adept at crowdsourcing or social networks. But perhaps it need not be so binary. ESRI's new ArcExplorer software (a free download) is their version of Google Earth and it can import both Google files and ArcMap shapefiles. Perhaps if ArcMap ever permits files to be exported into Google we'll see more integration of big GIS with the geoweb.

"The Democratization of Cartography"

"It turns out that when we talk about 'the world's information,' we mean geography too".

Google

Up until the 1980s it had always been assumed that maps were essentially devices that communicated information that had been gathered and processed by the expert cartographer. As the historical examples cited above testify, this had been the case for hundreds of years. The craft of cartography had a guild-like status; requiring years of training and the mastery of specialized techniques. These ideas about how maps worked were formalized in the post-war years by Arthur Robinson, a professor of geography at Wisconsin-Madison. Robinson provided the conceptual apparatus of what later became known as the Map Communication Model (MCM), which explains mapping as a process of communicating information from the map expert or cartographer to the map reader. The information is acquired, marshaled, and selected by the map expert and set down on the map.

As Mark Harrower, a leading proponent of populist cartography has observed:

> One of the themes of my profession right now is the *democratization of cartography*. ... Mapping used to be a job of the elite, the Rand McNallys and National Geographics of the world. Now people are taking it upon themselves to map their passions. (Science Daily 2006: n.p., emphasis added)

In other words, desktop mapping and geovisualization provided the beginnings of new forms of people's mapping. But the true democratization of cartography would only arrive with the advent of new advances in web technology, often referred to as Web 2.0 functionality such as massively distributed and hyperlinked datasets, mashups, and customizable open-source tools. These tools are profoundly different from their precursors because they allow *collaboratively linked* mappings.

Conclusion: Can Peasants Map?

Many of the obstacles such as the digital divide and net neutrality are not at base technological issues that can be addressed through market incentives, rather they are complex socio-political problems. Lack of access to online information parallels the very underserved populations it could be benefit. Community and participatory GIS, the netroots, and web-based mapping are therefore not likely to provide solutions for underserved populations to bootstrap themselves out of poverty. But if underserved and well-served communities work together then problems can be more ably addressed. This is a big if and as this chapter shows there are enduring divides *and* connectivities. That is, after all, because we live not in isolated communities, but in a world of networks.

In his work on political net-based activism, David Perlmutter explores the question of whether the online activism and the netroots are a representative constituency, specifically whether bloggers are "the people" (Perlmutter 2006). He points out that at the moment, the netroots are overwhelmingly young, white, male, well-educated, and tech-savvy and are thus not representative of the population as a whole. As he put it "peasants do not blog." In this chapter I have introduced a number of developments that both assist and create obstacles for access and usage of geospatial information. These tools are provided out of a genuine realization that the ways we visualize and understand the world around us – its places, geographies, and relationships – are undergoing a radical transformation. If the media (TV, newspapers, and news radio) has had to adapt and incorporate new models of information dissemination and participation, and if publishing is undergoing a similar transformation, then there would seem to be an equivalent transformation working on our mappings.

The remaining questions however are to what degree, how much, and with what effects these tools will confront the obstacles and barriers. The answers to those questions will prove vital in deciding the future of information.

Notes

1   In a sense mapping is never completed because things change. If mapping could be completed then the OS itself would have gone out of business after it finished its national topographical map series in the early twentieth century.

2   http://www.mydd.com/story/2006/10/9/232648/805.

3   http://www.dailykos.com/storyonly/2008/1/10/234313/397. On the question of solo vs. community blogging, almost all the progressive blogs have transitioned from the former to the latter. The "Scoop" technology, a so-called "collaborative media application" introduced in 2003, allows a site to be community driven by submitting stories, comments, and also act as editors and site managers. It is free under the GNU General Public License. In the words of one blogger "The days of the major, solo content generator, pundit blogger are all but over" (http://www.mydd.com/story/2007/2/6/142748/3955).

4   The FairData/FairPlan site is so vast that no description can really encompass it. It provides interactive maps, census data, precinct maps, or registered non-voters by race, racial profiling data, get out the vote data, and much more.

Chapter 4

What Is Critical Cartography and GIS?

"Cartography is not what cartographers tell us it is."　　　Brian Harley

Brian Harley's axiom quoted above that cartography is not what cartographers tell us it is can well serve to summarize some of the essential ideas behind critical cartography and GIS. Harley's life and contributions are examined in more depth later in this book (see Chapter 7), and his name is often invoked in the context of critical mapping. Calling things into question was a hallmark of Harley's life. One of his obituaries went to so far as to title itself "questioning maps, questioning cartography, questioning cartographers" (Edney 1992). How can his work help us understand the impact of critical cartography?

I would suggest that a good definition of critical mapping is that it *calls things into question*. Among the most notable things to be questioned are the claims of the discipline of cartography to be a science, and to progress from the solution of one problem after another in what Arthur Robinson called "The Essential Cartographic Process" (Wood and Krygier 2009).

A related idea is that critical mapping (cartography and GIS) examines the relationship of knowledge with power. What are the underpinning assumptions that help to govern knowledge? That is, what rationalities are in play? The reason many critical mappers and critical geographers think this is important is because these rationalities shape and form the subject of the map, that is, how the map helps oppress, subjugate, or subjectify individuals and populations (Wood and Krygier 2009).

To look at the relationship of power and knowledge therefore is *not* to claim that "knowledge is power" or that might makes right. What it does say is that what we know is affected by relationships of power: some ways of knowing are deemed to be better than other ways of knowing, and therefore it is "easier" for us to know things in certain modes rather than others. Which ways? Well, it depends on what historical time period you're looking at. Today, the scientific mode of knowledge is predominant. For a critical mapper, the objective is not to over-turn this way of knowing (as some scientists often believe) but to ask how it has come to be so powerful (perhaps as a historical investigation) and to ask what the implications are of this knowledge and whether or not alternative ways of knowing are possible. Because the latter question is sometimes framed as a critique of the limitations of scientific knowledge, or of its negative effects, some writers who identify with the scientific mode of knowledge have assumed that this kind of critique will usher in relativism. By this they claim that all ideas will become relatively acceptable; opening the flood-gates to non-scientific knowledge such as creationism, intelligent design, or worse, to the politicization of knowledge (e.g., to the denial of global climate change, or opposition to experimental stem cell research, etc.).

These disputes are long-standing and will not be resolved here. One point to bear in mind however is the understanding of critical researchers that knowledge can never come in an unpoliticized form, because as mentioned above they see know-ledge as situated within relationships of power.

It is significant that the word "discipline" has more than one meaning. In addition to referring to a body of knowledge such as geography, it also means the practice of learning (a related word is "pupil") and from that idea keeping order and control – in other words, power. Such order and control is what critical mapping attempts to deconstruct – in what way is it ordered? For whose benefit? Is it possible to con-ceive of mappings that are outside the control of the prevailing discipline?

So there appears to be a basic contrast between critique, which tries to open things up, and disciplinary knowledge, which tries to anchor and tie down. As we shall see in the last chapter, this divergence of effort results in one of the basic tensions that currently occupy cartography and GIS.

Undisciplining Mapping

In the last few years cartography has been slipping from the control of the powerful elites that have exercised dominance over it for several hundred years. You have probably already have noticed this with the emergence of fantastically popular mapping applications such as Google Earth. The elites – the map experts, the great map houses of the West, national and local governments, the major mapping and GIS companies, and to a lesser extent academics – have been confronted by two important developments that threaten to undermine their dominance. First, as Google Earth has shown, the actual business of mapmaking, of collecting spatial data and mapping it out, is passing out of the hands of the experts. The ability to make a map, even a stunning interactive 3D map, is now available to anyone with a home com-puter and a broadband internet connection. Cartography's latest "technological transition" (Monmonier 1985; Perkins 2003) is not only a technological question but a mixture of "open source" collaborative tools, mobile mapping applications, and the geospatial web.

While this trend has been apparent to industry insiders for some time, a second challenge has also been issued. This is a social theoretic critique that is challenging the way we have thought about mapping in the post-war era. During the last 50 years or so cartography and GIS have very much aspired to push maps as factual scientific documents. Critical cartography and GIS however conceives of mapping as embedded in specific *relations of power*. That is, mapping is involved in *what* we choose to represent, *how* we choose to represent objects such as people and things, and *what* decisions are made with those representations. In other words, mapping is in and of itself a political process. And it is a political process in which increasing numbers of people are participating. If the map is a specific set of power/knowledge claims, then not only the state and the elites but the rest of us too could make competing and equally powerful claims (Wood 1992).

This one–two punch – a pervasive set of imaginative mapping practices and a critique highlighting the politics of mapping – has "undisciplined" cartography. That is, these two trends challenge the established cartographic disciplinary methods and practices. It has certainly not occurred without opposition or resistance – which all new ideas encounter. For example, there is quite a strong trend in the USA and other countries right now to make people "qualify" as GIS experts through a licensing or certification process. Indeed an organization known as Management Association of Private Photogrammetric Surveyors (MAPPS) which represents licensed surveyors recently sued the US government in order to force it to hire only licensed users of geospatial information. This would have had large repercussions on federal contractors and further encouraged the development of "bodies of knowledge" that people must qualify in before they can use maps or GIS (such as this one: DiBiase et al. 2006). While MAPPS lost their lawsuit they issued a statement saying "the game is not over" (MAPPS 2007).

Critical mapping operates from the ground up in a diffuse manner without top-down control and doesn't need the approval of experts in order to flourish. It is a movement that is ongoing whether or not the *academic* discipline of cartography is involved (D. Wood 2003). It is in this sense that cartography is being freed from the confines of the academy and opened up to the people.

This chapter discusses these two critical movements of cartography. I will begin by elaborating on the idea that critique is political by its nature, which will sustain the view offered throughout this book that mapping is a political activity in and of itself. Second I will examine the idea that today's critical cartography and GIS, although it is often thought of as arising in the late 1980s associated with the work of Harley, Pickles, Wood, and others, is actually part of a longer-standing cartographic critique. Throughout its history mapping has been continually contested. Its history is not one in which it progressed from one stage to another as it became a science (an idea often implied in the textbooks). In fact, cartography as a way of knowing the world has constantly struggled with the status of that knowledge in a "contested tradition" similar to that of the geographical discipline more widely (Livingstone 1992a). As Brian Harley pointed out in his seminal article that helped launch the modern field of critical cartography, cartography is not merely to "what *cartographers* tell us maps are supposed to be" (Harley 1989a: 1).

What Is a Map? Why We Can't Define It and Why It Doesn't Matter

If cartography is not what cartographers tell us maps are supposed to be, then just what is a map? A typical official definition of the map is that "it is a graphic representation of space" (International Cartographic Association, ICA). That definition is fine as far as it goes, but it tells us little about the way maps are used.

If critique examines the assumptions of a field of knowledge, as we saw in the previous chapter, then it would seem that a fundamental critique of mapping is to carefully examine this definition.

For instance, is the concept of the map one which is invariant across cultures, age, sex, etc.? In a simple experiment that Roger Downs performed in his introductory human geography class, he was able to show some very interesting things about maps. Downs showed a number of images to his class (say 40 different pictures), composed of images taken from geography textbooks, aerial photographs, historic maps, and so on. For each image, you had to respond "yes it is a map," "no it is not a map," or "? Don't know."

Each person viewing the image made a personal assessment as to whether it was a map or not. The results showed the following:

1. in a given group, some images were almost always seen as maps, some were almost always not seen as maps, and some images were sometimes seen as maps;
2. the degree to which people agreed that an image was a map increased with age.

The first finding shows importantly that people have a core idea of what a map is (typically the kind of small-scale or global map found in textbooks). These are the maps that come to mind when someone talks about mapping or GIS. There are also a number of images that are almost universally rejected as maps (such as aerial photographs and often historical maps). So in any given group there is a lot of agreement on what constitutes a map and what doesn't and this covers most of the images. However, there is also a small number of images for which there is no agreement, neither being rated consistently as maps or not as maps by the group.

This second finding strongly indicates that our understanding of maps is a learned response, because as we gain more experience with the wide variety of mapping forms we are more willing to see them as maps. Children have a very narrow conception of what a map is, due to their narrower experience and understanding of symbolic representation. And while children begin remarkably early on to get a sense of how maps work, this sense remains subject to confusion and is incomplete for many years. For example, children often mix up scale (claiming to see the coach on the dugout of an aerial photograph of a baseball diamond), or are unable to separate the map symbol from its real life object (thinking that the road must be red because it is colored red on the map). As we get older the concept of the map correspondingly expands, although skill with mapping varies individually (Downs 1994; Liben and Downs 1989).

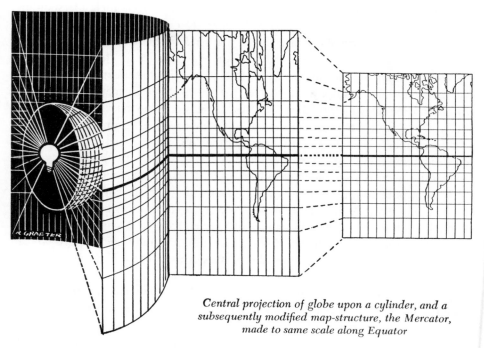

*Central projection of globe upon a cylinder, and a
subsequently modified map-structure, the Mercator,
made to same scale along Equator*

Figure 4.1 The shadow metaphor explanation of projections. Source: Greenhood (1964). © 1964 by the University of Chicago. Used with permission.

Even some skills we might expect to have attained by adulthood, however, remain elusive for most of us. Let's take an everyday example. One of the common ways to explain how a projection works is to ask the reader to imagine that there is a strong light source at the center of the earth. This source then "projects" a shadow in the shape of the landmasses on a container surrounding the earth (a cylinder for example). This explanation was once quite popular in cartography textbooks as Figure 4.1 shows.

However, when our actual understanding of shadows is tested we do quite poorly. Downs and Liben (1991) asked subjects (adult college students) to draw the shadow that would be cast by simple objects such as circles and squares. For example, if you cut a square out of a piece of cardboard and hold it between a screen and a light source, what would the shadow look like? What would it look like if you tilted the square? The results are startling; overall only about 50 percent of the shapes were drawn correctly, and this despite the fact that we encounter shadows on a daily basis in our lives. When the shape was held straight up, that is, at 90 degrees to the light source, the correct shadow was guessed more frequently, but when the shape was turned obliquely performance declined rapidly. Objects with thicker edges were almost never correctly drawn. The researchers also found that females did significantly more poorly than males.

If we want to answer the question of what is a map then we must begin by acknowledging that it is a culturally learned knowledge. Not only that but the skills necessary to comprehend them are learned through a *struggle* for understanding. Liben and Downs (1989) use the word "realize" to capture this sense of struggle; realize in the sense of making real, that is maps realize the world, and also realize in the sense of gaining an understanding ("Ah! Now I realize."). They contrast their position that maps are "opaque" in these senses of realizing to the commonly accepted one that maps are "transparent," that is that we see through maps to the underlying landscape (Downs and Liben 1988), or that it is possible to learn mapping early and easily. We can extend this argument to GIS: do we see "through" a GIS to the underlying reality (Downs 1997)? No, rather the GIS is a process of making a world (creating knowledge) not a mirror or window: "maps are creative statements about the world, not merely degraded reflections of it" (Liben and Downs 1989: 148). As the Iranian poet and mystic Rumi says: "speak a new language, so that the world will be a new world." This observation marks a key idea in this book.

So although people have a strong idea of a typical map (the "core" map concept) the kinds of maps that comprise this core are not culturally invariant, as we can quickly see if we look at maps outside the Western tradition. Figure 4.2 shows a native American map and a Pre-Conquest Mixtec map.

Speaking of maps such as the Nuttall Screenfold, Harley observes that it "does not look like a map to twentieth-century eyes. Yet as a picture history telling the story of an early conqueror of Southern Mexico it is fixed in space as much as time. When we crack the code it reveals elements that are map-like in purpose and content" (Harley 1990b: 29).

The conception of a map varies significantly between different cultural groups. The map test also found a large number of images that are often but not always understood as maps, perhaps depending on the context. The degree to which these map-like objects are rated as maps increases as people become more familiar with them. Maps appear to exist on a sliding scale of "mappiness," varying from extremely mappy to only slightly mappy. There is however no single essential "look of maps" – this is why people have struggled with an all-encompassing definition and why it is more productive not to try (Vasiliev et al. 1990). They can and have appeared on bark, animal skins, papyrus, linen, paper, clay, wood, sand, rock, computer screens, napkins, and backs of envelopes, to name a few physical media. This does not include mental maps nor maps that only exist as performances (for example a policeman gesturing a route or an artistic performance of a map – see Chapter 12).

Maps then are part of the cultural knowledge that we acquire by being immersed in a society. Both our expectations *about* maps (what they should look like, how to use them) and the play of knowledge that they *produce* are deeply related to the shape of that culture and its contours of power.

The Production of Space

Consider the word "representation" in the ICA definition. Critical cartography and GIS question what is meant by "representation," a question that has also often popped

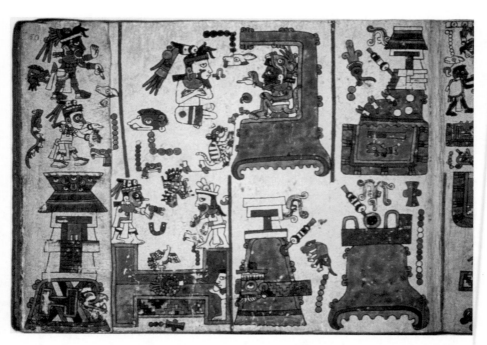

LIBRARY, UNIVERSITY OF CHESTER

Figure 4.2 Ojibwe (Native American) map, c. 1820, detail (top) and Nuttall Screenfold Pre-Conquest Mixtec map (bottom). Source: Ojibwe map drawn by John Krygier, used with permission. Nuttall Screenfold © Trustees of the British Museum, used with permission.

up in philosophy (Rorty 1979) and cultural studies, as well as geography (Thrift 2006). This is because "representation" naturally enough appears to imply that something already exists prior to the act of mapping (the space or landscape being mapped). There is a landscape out there, and it is captured in some "representative" way by the map. Even if we agree that the landscape is not the map, maps have to creatively leave out details as Monmonier has written: "not only is it easy to lie with maps, it's essential" (Monmonier 1991: 1). But still, the landscape comes first, and like a painting or photograph we "take" essential elements of it for our representation.

Critical cartographers (as well as cognitive developmentalists such as Roger Downs and Lynn Liben cited above), on the other hand, argue that mapping creates specific spatial knowledges and meanings by identifying, naming, categorizing, excluding, and ordering. The ICA definition of the map as a graphic representation does not exclude this meaning, but it doesn't really emphasize it either. Furthermore, once these categories are put into play they can be used to exert power and control people and things. Mapping creates knowledge as much as (and for some, instead of) reflecting it. Critical cartographers do not argue that *physical* space is produced by the process of mapping, but rather that new ways of thinking about and treating space are produced. "Space," in this account, is not just a question of physical and material disposition (although it is that) but also the constitution of objects. For critical cartography, mapping is not just a reflection of reality, but the production of knowledge, and therefore, truth.

For example, when Christopher Columbus struck land on October 12, 1492, he carried with him (or had seen) a map or globe much like the one reproduced in Figure 4.3, which shows a facsimile of the Behaim Globe.

Figure 4.3   The Martin Behaim Globe (1492). From the American Geographical Society Library, University of Wisconsin-Milwaukee Libraries.

Columbus of course was unaware of the American continent, but contrary to popular mythology he knew the world was round. He drew on the work of ancient geographical writing by Aristotle and Ptolemy who themselves had tried to estimate the size of the globe, and had mapped the known portions of it. Columbus's plan was to sail westwards to India and China to establish a new trade route but also to convert the inhabitants to Catholicism for King Ferdinand and Queen Isabella (who were financing his trip). He named his journey the "Enterprise of the Indies" and attached to it all sorts of titles and land bequests for himself and his family. As Harley has shown (1992b), he renamed with Christianized names the islands and places already named by the indigenous Sarawak Indians. So the place where he made landfall became "San Salvador" (the Savior), other islands became Santa Maria de la Concepcíon (holy conception), Trinidad (holy trinity), etc. Even his ship was called the *Santa Maria* (Virgin Mary) and he signed himself "Christoferens" or "Christ-bearer." In fact he succeeded so well in renaming places that only one original name is known, the island of Guanahani (an island in the Bahamas). Columbus's cartographer, Juan de la Cosa, made an interesting map of these discoveries, placing the new, religious names on the map (Figure 4.4).

By reinscribing new identities on these places then, and specifically with Western Christian names, Columbus effectively created a new space that was compliant with Western beliefs, and which permitted it to be governed and controlled. As Harley observed:

> The purpose of [the Juan de la Cosa] map is clearly signposted as an instrument of European empire. National flags – both Spanish and English – are planted to claim ownership of the new territories. The map also proclaims a crusade. A compass rose astride the equator portrays the Holy Family. The figure of St. Christopher is said

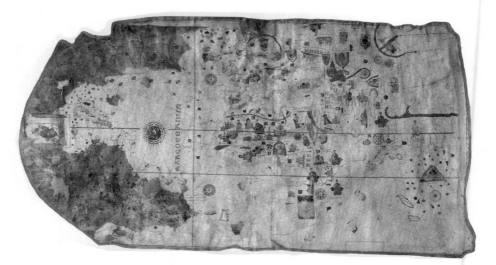

Figure 4.4 Juan de la Cosa's map of the world, *c.* 1500. Columbus' discoveries are indicated on the left-hand side.

to be an allusion to Columbus carrying the Christ child on his shoulders. As "Christoferens," he is the bearer of Christianity across the ocean to the pagan shores of the New World.

Placenames commemorate the famous shrines of the Virgin in Castile, Catalonia, and Italy. Placed thus on the new land they become emblems of possession. Columbus tells us in his Journal for Friday, 16th November 1492: "in every place I have entered, islands and lands, I have always planted a cross." The names on the map are the written record of these innumerable acts of territorial consecration, some of them witnessed by Juan de la Cosa. (Harley 1990b: 61)

This is a classic episode in the history of cartography and colonialism. It demonstrates that maps make space as much as they record space. This is quite literally "map or be mapped" (Bryan 2009; Stone 1998). As Bernard Nietschmann once pointed out "more indigenous territory has been claimed by maps than by guns," but as we will see in more detail in Chapter 9, the corollary of this is that "more indigenous territory can be reclaimed by maps than by guns" (Nietschmann 1995: 37).

In the next chapter, we will delve further into this assertion that mapping became scientific during the second half of the twentieth century. On what basis is this claim made? Why is it made? In subsequent chapters we shall expand the scope of what mapping does, specifically we shall see how maps are involved in governance, in geosurveillance, and in the construction of race and identity.

Chapter 5

How Mapping Became Scientific

Countable Information

Around the time of the Second World War a rather unorthodox engineer could be seen coming in to work at the Bell Laboratories in Florham Park, New Jersey (now the AT&T Labs). By the end of the war, Claude Shannon had already been working at the Labs for seven years but was still only 32 years old. Shannon was the son of a probate judge and a language teacher and he had a playful, eccentric personality (Gleick 2001). He was often seen riding through the hallways on a unicycle of his own design while juggling three balls. Shannon would have fit in perfectly with the exuberant dot.com boom of the late 1990s and early 2000s where the young dotcommers rode through their offices on motorized scooters or Segways.

Shannon loved juggling and even created a tiny mechanical stage on which three clowns juggle a number of rings and balls. He also designed and built chess-playing machines. Shannon also had a kind of maze-solving device incorporating an electromechanical "mouse" named Theseus, which was one of the earliest experiments in teaching a machine how to "think" (Bell Labs 2001). Although these demonstrate a whimsical side to his nature they were all directed to some end. The unicycle for example was a practical experiment in control under conditions of inherent instability.

Shannon joined the Bell Labs in 1941, where he worked on cryptography. One theorem of Shannon's was used in the original "hotline" between the UK and USA. Known as the "SIGSALY" voice encryption system, it allowed President Roosevelt to speak with Winston Churchill. The device weighed over 50 tons, and consisted of over 40 racks of equipment and was effectively limited to major world capitals, although one was also installed on a ship which trailed General Douglas MacArthur around the Pacific. The London terminal was installed in the basement of Selfridge's Oxford Street department store, near the Admiralty War Room. A modern reproduction of the SIGSALY can be viewed in the National Cryptologic Museum (Fort Meade,

MD). The system is significant because it was one of the first to use information encoding methods.

Both Shannon and Norbert Weiner worked on control systems or cybernetics to develop a better firing system for anti-aircraft systems. "Cyber" means "controller" or "governor" in Greek (literally the steersman of a ship). Their work was funded by the government (Vannevar Bush's National Defense Research Committee, NDRC) to the tune of about $10 million spread over 80 different projects (Mindell et al. 2003). Target acquisition was a very difficult problem at the time; the planes moved fast and not always in predictable ways, there was an unavoidable delay between firing the missile and it reaching its target (sometimes as long as 60 seconds) during which time the aircraft could alter course, and it was also difficult for the control mechanisms – the servos – to acquire and hold the precise positions needed to accurately aim the missiles. And of course the human operators could make mistakes or fire too late. The most promising approach was to model the world as an information system. Not only could the inputs be represented as signals, but so too could the performance of the fire control system, that is, by using a special feedback loop (one of the first times this method had been used) the servo could monitor its position and accurately maintain the correct position. As one excited journalist put it:

> Those machines are designed to collect and elaborate information in order to produce results which can lead to decisions as well as to knowledge . . . [it is] a unique government for the planet [which could lead to a new] political Leviathan. (quoted in Mindell et al. 2003: 75–6)

The rhetorical appeal to Hobbes's Leviathan was not accidental. The recognition of the political application of information-as-knowledge, while it exceeded any of the plans that Shannon had for his ideas, was a timely reminder that knowledge is politics. More precisely it was a realization that technology was an arm of neoliberal governance, quite literally "cyber" or control technologies. In France in 1947 (where Weiner's book on cybernetics was first published) there were strikes which caused the cancellation of one of the cybernetics lectures, and a general feeling of antipathy to the United States. But cybernetics could supposedly unify science and culture, for its basis as a system of control could be applied to the populace as much as to firing mechanisms.

An interesting parallel is provided by the early career of Waldo Tobler, familiar to most geographers today through his First Law of Geography. In the late 1950s, Tobler worked on SAGE (Semi-Automated Ground System) at a RAND Corporation spin-off in Santa Monica, California. SAGE was an automated tracking system designed to monitor Soviet nuclear missiles and bombers. In this respect Tobler "was one of the first in geography to think through the possibilities of bringing together computers and cartography" (Barnes 2008: 12).

Today, Shannon is credited as being one of the greatest scientists of the twentieth century. Much of the reason for his achievement rests on his invention of communication theory, or what we would call information theory today. This theory is at the heart of our digital devices such as computers.

Shannon's achievement was that he recognized that information was "countable." He was able to define information in a way that made it useful for all sorts of communication devices (such as telephones, but also the Internet and the Web). Using his methods it was now possible to count the maximum amount of information that it was possible to transmit through a particular channel (Mindell et al. 2003).

What does it mean that information is countable? Just what is a "unit" of information? Or to put it another way, what is the smallest amount of information you can have? Although you might think the minimum is "one" the answer is actually "two." The reason is that one of something doesn't allow you to differentiate. Shannon defined information as that which allows you to make a decision. For this you have to be able to differentiate between at least two possible states.

Let's say you have a condemned man about to die and you are waiting for a phone call from the governor on whether he will commute the sentence or not. There are two possible messages that you could get when the phone rings: a "yes" or a "no." That's two possible states. If the phone rang but was silent you wouldn't know what it meant. Another example would be heads or tails of an ordinary coin. No one says "I'll decide whether to do something whether it's heads *or* tails"! Using these ideas, Shannon defined information as that which allows you to make a decision (execute or not execute the prisoner) and the basis of information as the *binary* digit, or "bit." Shannon first used this word in 1948, attributing it to John Tukey. Over time, these amounts of information have been given names. Notice that they progress as powers of 2:

$2^1 = 1$ bit
$2^2 = 4$ bits
$2^3 = 8$ bits = 1 byte
$2^{10} = 1,024$ bytes = 1 kilobyte
$2^{20} = 1,048,576 = 1$ megabyte
$2^{30} = 1,073,741,824$ bytes = 1 gigabyte
$2^{40} = 1,099,511,627,776$ bytes = 1 terabyte
$2^{50} = 1,125,899,906,842,624$ bytes = 1 petabyte
$2^{60} = 1,152,921,504,606,846,976 = 1$ exabyte (a billion gigabytes!)

These are the ones in most common usage (for example in measuring the amount of RAM memory on your computer – hard-disk space is usually measured in decimal, not binary units). Under Shannon's system we can measure the size of the hard drive, the amount of memory (RAM), or the throughput of your internet connection. The latter is know as the "bit rate" and represents the number of bits that pass through in a second. Early modems for example operated at 14.4 bits per second.

One exabyte represents a tremendous amount of information. Amazingly, attempts have actually been made to calculate the total amount of information produced each year! The answer: 5 exabytes (Lyman and Varian 2003). Frankly this is just a nice guess. Another more recent attempt estimated that the "digital universe" in 2007 of information created (and replicated) was actually closer to 281 exabytes, and could be ten times that size by 2011 (Gantz 2008). Whatever the size, there's no doubt that there is a lot of information around (and more arriving every moment). Some people are beginning to talk about an "information footprint" analogous to the carbon footprint: how much information you "shed" or produce each year. Early experiments with "lifelogs" or the practice of completely logging your own life (including 24/7 audiovisual data capture) indicate it could be many, many gigabytes per year (Dodge and Kitchin 2007; A. Wilkinson 2007).

Shannon's breakthrough paper came one day in 1948, shortly after the war. It was entitled "A Mathematical Theory of Communication" and was published in the Bell Labs own technical journal (Shannon 1948). In addition to defining and counting information, its key ideas were that communication could be improved if the "signal" (the information) was maximized and the "noise" (the unwarranted distortions or errors) could be minimized.

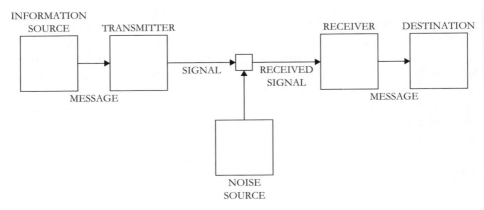

Figure 5.1   Shannon's 1948 schematic of a general communication system.
Source: Redrawn from Shannon (1948).

Working in the context of telephones, Shannon's idea was that a signal was trans-mitted at one end, passed through a communication medium or "channel" (the phone wires and equipment) and received at the other end. In order to achieve this, Shannon demonstrated another key idea concerning information: it was *fungible*, that is, it could be converted into different forms. A voice for example, could be converted into ones and zeros and reconverted at the other end, and yet despite its digital nature it would still sound like the original! This is a deeply radical idea, because it means that any object, whether it be a Renaissance painting, a sound, or a photograph could be converted and transmitted over distance. Or, of course, a map.

During the second half of the twentieth century, this "communication model" was freely adopted in cartography to explain how maps work. One of the leading proponents was Arthur Robinson (Robinson 1952), and it is his contribution perhaps more than any other that characterizes the nature of cartography during the second half of the twentieth century, and thus the story of how modern carto-graphy became scientific.

Arthur Robinson and the OSS

In order to understand how ideas from information theory could be so readily adopted in cartographic theory during the late 1960s, we need to do a little excavation of ideas that were set in place by Arthur Robinson (1915–2004).

Despite Robinson's obvious influence on the field his career remains relatively unexamined, and somewhat surprisingly there are no collections of his works or biographies about him (Edney 2005b). Yet it would be difficult to understand where he's coming from without knowing a little of his life.

Robinson was recruited by the geographer Richard Hartshorne in 1941 to join the geography division of a nascent intelligence service, the Office of Strategic Services (the OSS, which would later become the CIA). As many as 200 geographers may have worked for the OSS (Barnes claims 129, see Barnes 2006; Stone 1979). Hartshorne, who by that time was a major figure in geography having published his definitive manifesto *The Nature of Geography* in 1939 (Hartshorne 1939), was Director of the Projects Committee of the OSS. This committee was housed within the OSS Research and Analysis (R&A) Branch. As its name implies this branch was responsible for gathering and assessing intelligence and it was here that the important Map Division was established, headed by Robinson. Besides Robinson, the R&A Branch also included the geographers Preston E. James, Kirk Stone, Chauncy Harris, and Edward Ullman, as well as other intellectuals such as Walter Rostow, Arthur Schlesinger, Jr., and Herbert Marcuse (Barnes 2006; Katz 1989). Effectively it was an organized collection of academics in service of United States wartime intelligence, and as such is a direct descendent of the "Inquiry" of World War I established by President Wilson (Crampton 2006; 2007b).

In 1942 Robinson's position was Chief of the Map Division at the OSS, which in its short lifetime made 8,200 new maps (including huge 50-inch globes that were given to both Winston Churchill and President Roosevelt (Robinson 1997),[1] answered 50,000 requests for map information, and provided cartography for four Roosevelt–Churchill Allied Conferences (Barnes 2006). According to one of his contemporaries, the geographer Lawrence Martin, "during the fighting days it was [Robinson's division] and not the Army's Military Intelligence . . . who made the maps for the Joint Chiefs of Staff" (Martin 1946/2005).

In January 1946 Robinson started work at the University of Madison-Wisconsin and obtained his PhD in 1948 entitled the *Foundations of Cartographic Methodology*. In a 1997 interview he talked about how he got interested in this subject.

> I remember picking up books having to do with art and color and the graphic arts and so on. They opened my eyes to the fact that there had been a lot of psychological work going on, perception of lettering and things like that. It all seemed to fit very well with map making. (Cook 2005: 49)

His war work has often been noted as the impetus that drove him to call for scientific research on map design (Robinson 1979; 1991; Robinson et al. 1977).

Perhaps Robinson's best known contribution is his development of the map as a communication system. This focus had the goal of improving the efficiency and functionality of maps as communication devices via empirical experimentation.

Robinson's job at the OSS was to provide unbiased and reliable maps of the military theaters and landing zones. At a time of increased cartographic propaganda by both sides – Nazi maps claiming to show Germany surrounded by enemies for example – Robinson wanted to ensure that map design was clear, efficient, and effective (Edney 2005b). The OSS map unit was given official status in the European theater of operations. The OSS microfilmed the map archives of the Library of Congress, the Department of State, and the American Geographical Society but suitable maps were in such short supply that the Director of the OSS William "Wild Bill" Donovan put out a public radio appeal (Wilson 1949). There was even a formal contract with Arnold Toynbee, then the Chief of the British Foreign Office Research Department, to provide British political maps, and OSS advance map teams were even in the Justus Perthes map plant in Gotha, Germany, while it was still partially occupied by the Germans (Wilson 1949)! The US Army Corps of Engineers performed similar actions, perhaps most notably by Edward B. Espenshade who for many years edited the well-known Goode's atlas (given to generations of American freshmen), and who was an intelligence specialist who entered liberated cities during the war to secure mapping and aerial photo intelligence materials.[2]

As we might appreciate, there is a deep and long-abiding connection between military wartime intelligence and the production of geographic and cartographic knowledge. Robinson's experience and career (and he was far from alone) is an exemplar of this for WWII, as was the OSS more broadly, and, during WWI, the Inquiry. A critical GIS and cartography however must also concern itself with "governmental" production of geographical knowledge outside times of "official" war, or during quasi-wars such as the War on Terror. The lesson of the OSS then is not that it was a historical event, but that it produced particular kinds of geographic knowledge under particular circumstances – and the same is true today. "War science," as it has been called (Barnes and Farish 2006), will have effects after war is over and the political state of "exception" need not be temporary (Agamben 2005).

In the next section we can trace more closely how this experience directed Robinson's influence on the field of mapping.

From OSS to a Scientific Discipline of Cartography

After the war, Robinson started to distance cartography from art and design. How was this to happen? For Robinson "function provides the basis for design" (1952: 13), that is, *form follows function*. He found indications that the field was already moving to this "functional" approach, identifying a study in 1933 as the first to take this approach:

> Cartography is neither an experimental science in the sense that chemistry or physics are nor is it searching for truth in the manner of the social sciences. Nevertheless, it employs the scientific method in the form of reason and logic. (Robinson 1953: 11)

It is disconcerting, Robinson argued, to rely on "subjective artistic or aesthetic" sensibilities in map design, and if creativity and imagination did have a role in mapping "it is equally important that significant processes be objectively investigated [in] the visual consumption of a graphic technique" (1952: 17). A "nicely drawn" map is not necessarily a "good map" (Robinson 1953: 9). Function should operate even in the aesthetic components of the map.

Robinson suggested that there were two routes open to achieving his desired binary division of scientific from nonscientific cartography. One was to completely standardize everything ("and the absurdity of such a proposition is, I hope, obvious" [Robinson 1952: 19]),[3] the other is to analyze how people actually use maps. For Robinson this meant studies of perception. He appealed to other disciplines where he found the mechanisms to pursue these studies, and in a famous formulation laid these out:

> the development of design principles based on objective visual tests, experience, and logic; the pursuit of research in the physiological and psychological effects of color; and investigations in perceptibility and readability in typography. (Robinson 1952: 13)

This statement was to indicate a new direction in cartography, that of perceptual and cognitive studies of mapping and map use that lasted well into the 1990s.

Robinson saved his strongest attack on artistic elements of mapping for those which "attempt to awaken various responses not necessarily of beauty" as he put it, that is, mapping for political propaganda purposes (1952: 18). Having just emerged from a global war in which propaganda maps were deployed extensively on both sides (for examples and reproductions, see Pickles 2004: 37–47), Robinson was understandably anxious about this aspect of mapping. He wanted to reject anything that was not part of the map's function or purpose, especially those aspects that seek to "unduly" sway the reader to some point of view about other people or countries. It was almost as if maps should behave with a sense of propriety and good manners, and that to transgress beyond these proper behaviors was worse than just an exercise in bad taste, it was also a sign of a bad map (Krygier 1996; Robinson 1952; Robinson and Petchenik 1976).

Robinson is often described as the "dean of modern cartography," a man who introduced the scientific method to cartography (Montello 2002). It is important to put this claim into context however, as cartographers had sought to formalize the discipline since at least the early 1900s and 1910s, if not before (Jefferson 1909; Wright 1930). It should not be forgotten either that the great surveys of the eighteenth and nineteenth centuries were carried out using the Enlightenment principles of rationality, not to mention trigonometry and mathematics. For example, the famous Cassini family was instrumental in surveying and mapping modern France (the *Cassini* spacecraft, which is the first to orbit Saturn, was named after the first Cassini). No less than four generations of Cassinis were involved in the first scientific map of France, which in some areas resulted in significant repositioning and resizing. Visiting their observatory in 1682 King Louis XIV is reputed to have said on seeing his new shrunken state "you have cost me more territory than all my enemies!" Beside this increased accuracy, the most remarkable achievement of the Cassinis was to bring together disparate local knowledges into one coherent knowledge base, their "Carte de Cassini" which was finally published in 1789.

A centralized knowledge base went hand in hand with the emerging modern political state, a unique system of measurement (the metric system), and a common set of instruments (a standard scale, quadrants, survey chains, and triangulations). Such a centralized, almost panoptic system, is characteristic of rational scientific knowledge-creation (Turnbull 2003).

Robinson was not alone. He worked within a specific intellectual-situated context – not only his personal one, but that of the time. Other influential work by early twentieth-century cartographers (and by this time it was indeed possible to speak of a discipline of cartography) that should also be recognized as influencing Robinson was Erwin Raisz, especially his textbook of 1938 (Raisz 1938). Raisz (1893–1968) was of Hungarian origin, learning cartography in Budapest, and he served in the defeated Austro-Hungarian army during World War I. In 1923 he emigrated to the United States and went to graduate school at Columbia and then was hired at Harvard. At the time both of these universities had well-known geography departments; Harvard had been the home of William Morris Davis and geography at this time had a strong physiographic influence. At Columbia Douglas Johnson (who had been deployed by the American government during the war to scope out the terrain of Europe for Military Intelligence [Johnson 1919]) was professor of physiography.

This suited Raisz because one of his most remarkable skills was the ability to draw landscapes by hand. These pen and ink landform maps are still available today. Raisz drew plenty of them – not just the United States but also South America, Europe, Africa, and Asia. When I was a student, one of the professors, Peirce Lewis, would enthusiastically advocate the Raisz landform map of the United States to his classes, calling it "the best map of the US irrespective of subject!" (P. Lewis 1992: 298).

Another Kind of Binary Mapping

What can Robinson's work tell us about the development of modern cartography? In establishing his axioms Robinson introduced for the first time the notion of the "proper map" and its opposite, the transgressive map. The proper normal map would be defined by a series of positive, scientifically derived statements and qualities (it should have this or that quality), while the transgressive map was merely defined as its opposite, a map that lacked these qualities. Rather than having other qualities, the transgressive map was a negative space, a void or lack. As such, this partitioning exhibits the classic properties of the binary divide, a way of thinking that has often pervaded the social sciences (Cloke and Johnston 2005).

These binary divisions are nonetheless common enough in geography (so much so that a recent book is dedicated to them [Cloke and Johnston 2005]). In 1989 Brian Harley identified a similar binary series that he thought was often applied to mapping as a "discourse of opposites":

Table 5.1   Harley's discourse of opposites.

"Artistic" Maps	*"Scientific" Maps*
Aesthetic	Non-aesthetic
Autographic	Anonymous
Imaginative	Factual
Subjective	Objective
Inaccurate	Accurate
Manual	Machine
Old	Modern
Place	Location

Source: adapted from Harley (1989b)

Harley argued that this discourse of opposites has detrimentally structured the history of cartography. He wrote that "the tendency of modern science – which has hijacked cartography – [is] to create a series of fundamental dualisms to define and defend its own territory. Classification is part of the way in which science claims to control and to mirror the world" (Harley 1989a: 6). While Harley is influenced by the work of Derrida in the same way as Gregory, he was more influenced by the work of art historians who insisted that mapping as a form of representation should be placed into its particular, local, and contingent context: "[m]aps are always a part of culture and never outside it" (Harley 1989a: 18).

This last sentiment – so close to Derrida's idea that there is nothing outside the text, i.e., that nothing escapes its place in a chain of signifiers – is a common feature of critical work and has led to an emphasis on *local knowledges* rather than global knowledge. Local knowledge emphasizes its situatedness (including its geographical situatedness). Local knowledges give partial perspectives (and have perspectives rather than God's-eye views), and are self-conscious of their position. Apparently then, Robinson's desire that maps operate outside of their political, social, or aesthetic contexts was always going to be wrong. Nevertheless, it has proven influential to several generations of writers on mapping and GIS and many traces of it can still be found today (see also our discussion of the great Peters map controversy).

Robinson's achievement was to break with previous conceptions of mapping in which the artistic or aesthetic component could sit alongside the scientific and to subsume the former into the latter. Mapping became the portrayal and communication of scientific data, and in this he set up the way for the development of the map communication model in the late 1960s. By 1972 this model was firmly established in the discipline, with the International Cartography Association (ICA) establishing "the theory of cartographic communication" as one of its terms of reference (Ratajski 1974: 140). Robinson secured this understanding in his influential book with Barbara Bartz Petchenik *The Nature of Maps* (Robinson and Petchenik 1976), a title which has been echoed (in ironic mode) by both Brian Harley, who needed a *new* nature of maps (Harley 2001), and by Denis Wood and John Fels, who needed new nature*s* (plural) of maps (Wood and Fels 2009).

Robinson and Petchenik argued that the proper study of cartography had only begun with the study of what they called the percipient, "that is, the receiver of the information prepared by the cartographer" in the 1940s (Robinson and Petchenik 1976: 24). "It was not until the development of formal communication theory and investigation [they mean Shannon] that cartography incorporated this aspect," they added (Robinson and Petchenik 1976: 24).

Writing in 1977 he could offer the following observation:

> The goal in 1950 was simply to make a map; in 1975, in theory, a map maker makes the map created by a cartographer who is supposed to be sensitive to the capabilities of his envisaged map reader. Corollaries of this view are a lessened concern for the map as a storage mechanism for spatial data and an increased concern for *the map as a medium of communication....* In communication the psychology of the map reader should set upper and lower bounds on the cartographer's freedom of design. (Robinson et al. 1977: 6, emphasis added)

The phrase "the map as a medium of communication" indicates that the map was primarily a means to communicate information and that it was necessary to take into account the information capacities of three components: the cartographer, the map reader, and the map itself. This informational capacity would drive map design.

The Map Communication Model (MCM)

There is a remarkable graphic published in a recent issue of *Cartographic Perspectives* that shows how deep Robinson's influence was on this generation of cartographers. The graphic shows, in an inner circle, Robinson's PhD students, and then, like throwing a stone into a pond, the ripple effect of those who were their students, and so on outwards. The graphic contains some 199 names. So Robinson's ideas were transmitted both directly through his students, and more widely still, through his textbook. It was not until the mid-1980s that *Elements of Cartography* was even seriously challenged in the field. For over 35 years Robinson's name was indissolubly linked to the academic pursuit of mapping. Robinson advised 15 PhD students, including several who went on to pursue careers in cartography such as Norman Thrower, Judy Olsen, Henry Castner, and David Woodward.

The map communication model reached its apotheosis in the late 1960s with an extremely complex version published by the Czech cartographer Koláčný (1969). This version incorporated multiple feedback loops to account for criticisms of the MCM, for example, that the user might give input to the cartographer after having used the map, for example if the users were national mapping agencies who had hired the cartographer. While it still clung doggedly to a separation of cartographer from user, it was becoming clear that the model was unwieldy and could mean almost anything as one traced its various loops and interconnections.

A simplified map communication model is shown here.

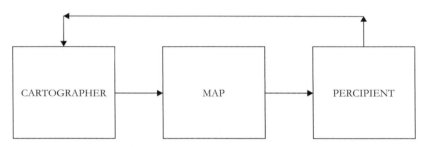

Figure 5.2 Simplified map communication model.

The message that the cartographer wishes to convey is encoded in the map, and it is here that Robinson's influence can be readily detected, for the MCM also included consideration of how that message was received. Previously, cartography stopped at map design without consideration of map use, perception, or recall. But now the map is a way-station through which the message travels from the cartographer to the map user (Robinson's "percipient").

One of the reasons for the turn to modeling in cartography was the dominance of spatial science in geography in the post-war period up to the 1960s. In the late 1960s this dominance was codified in Chorley and Haggett's influential book *Models in Geography* (Chorley and Haggett 1967). It is not surprising then that a major outline of the map communication model was included (Board 1967). Like Robinson almost two decades earlier, Board was concerned that mapping was governed by arbitrary guidelines, not only in map design, but in the very way that maps functioned. What was needed, he argued, was a more formal understanding, or model, of how maps work. Drawing from the great intellectual excitement in quantitative and modeling work then being done in geography (the quantitative revolution), as well as attempts to "mathematicize" cartography (Bunge 1966), Board outlined a processual model that encompassed not only the map, but data collection, cartographic generalization of data, design decisions, and the reception of the map by the user. This model conceived of mapping therefore *as a process*, in which information was gathered, shaped by the cartographer, encoded in the map, decoded by the viewer, and absorbed. It involved both intellectual-cognitive issues, perceptual issues, and questions of data management and representation.

Board was able to turn to Claude Shannon's outline of just such a process in the transmission of signals. He presented Shannon's generalized communication system as he had described it in Johnson and Klare (1961). Thus, where information theory spoke of "noise" or interference in the signal, so now could cartographers seek to identify obstacles to the successful transmission of geographic information from the cartographer to the user via the map. These obstacles could include poor map design, bad generalization decisions, or unsatisfactory map viewing conditions (for example under poor lighting, or while driving a car).

With the increasing development of computing power and availability at this time the MCM was also easy to integrate into an emphasis on technological developments. The 1960s and 1970s saw a number of significant developments in the technology of mapping and GIS. Two of today's most influential GIS companies were founded in 1969 (ESRI and Landscan), the Harvard Graphics Lab was in full swing producing early GIS software such as ODDYSEY (Chrisman 2006), and the Cold War was prompting the increased use not only of aerial photography but, after the American U-2 spyplane was shot down over the Soviet Union in 1962, the development of high-resolution satellite imagery. The US government also launched its highly successful "Landsat" series of satellites in the early 1970s which collected a vast array of environmental data, including images of the earth from space that helped generate a sense of the earth as a "blue marble" hanging in space (Cosgrove 2001).

The "Mangle" of Geographic Knowledge and Science

In this chapter we have traced part of the career of Arthur Robinson with a view to understanding how mapping became scientific. Of course, Robinson drew on his formative career experiences and on the work of earlier writers such as Claude Shannon. But the point to grasp here is that if mapping and GIS became more scientific they did so as practices of "government" during wartime and its aftermath. "Peace" geographies are not divorced from war geographies. Furthermore the science that is produced was twisted together with geopolitical aims. Trevor Barnes has called this "the mangle" (Barnes 2008). The mangle is how politics, military intelligence, and geographical knowledges are not produced in isolation from one another.

As Pickles points out, these kind of communication and information models were developed "at techno-scientific research labs such as those at Harvard, MIT, Berkeley and Bell Labs [Claude Shannon's employer], and those more directly funded by the US Government's combined efforts to . . . build Cold War security institutions" (Pickles 2004: 33).

While they were popular for a while, authors now identify Robinsonian studies as conducive of nothing less than an "ontological crisis" in cartography (Kitchin and Dodge 2007) because they imply an unwarranted ontological security. Maps have passed themselves off as natural representational devices (recall the "perverse sense of the unseemly" from Chapter 1). Kitchen and Dodge push for an under-standing of maps as "ontogenetic" or "of the moment" and constantly brought into being:

> Our thesis is that ontological security is maintained because the knowledge under-pinning cartography and map use is learned and constantly reaffirmed. A map is never a map with ontological security assumed; it is brought into the world and made to do work through practices such as recognizing, interpreting, translating, communicat-ing, and so on. It does not re-present the world or make the world (by shaping how we think about the world); it is a co-constitutive production between inscription, individual and world; a production that is constantly in motion, always seeking to appear ontologically secure. (Kitchin and Dodge 2007: 335)

In his efforts to provide a firm foundation for cartography then, Robinson seems to have engendered not a resolution, but "anxiety" and "crisis." In the next chapter we shall look more closely at this relationship between maps, power, and know-ledge by examining the roles of mapping in governance.

Notes

1 Part of a set of only a dozen custom-made globes. One found its way to the University of Georgia geography department, about 70 miles from where I write this, while others are in the private homes of Winston Churchill at Chartwell, the American Geographical Society Library, and the Library of Congress.
2 See Espenshade obituary www.northwestern.edu/newscenter/stories/2008/02/espenshade. html.
3 Although not to the previous owner of my copy of the book, the cartographer Borden Dent, who puts in the margin the single word "why?"

Chapter 6

Governing with Maps:
Cartographic Political Economy

"It escapes me how politics etc., can enter into it."
                              Duane Marble (cited in Wood [1992: 240])[1]

*"Maps are implicated in the exercise of power as political geographers are
well aware."*
                    Editor of the journal *Political Geography* following
                              Brian Harley's death (Taylor 1992)

Duane Marble's complaint, made in the context of the great controversy over the
Peters world map that rocked cartography for nearly two decades (see Chapter 7)
is emblematic of the assumption that maps are naturally a-political. Peter Taylor's
Political Geography editorial, working from a completely contrasting set of assumptions, is also emblematic. In this chapter we shall examine the merits of these
differing views and how maps and GIS might figure in the political economy.

If you ask the average person on the street if maps are political they might say "yes,
some maps are," or "yes, they *can* be, but this should be minimized!" In Chapter 5
we saw how this was the view of Arthur Robinson, who wanted to eliminate map user
responses "not necessarily of beauty," that is, the political messages carried by maps.
More recently, the popular writer and cartographer Mark Monmonier has said that
political mapping amounts to nothing more than drawing a few redistricting lines
(Monmonier 2002a). This is especially odd in his case because many of his books
explore the politics of maps, surveillance, and decision-making (e.g., Monmonier
1989; 1991; 1995; 1997; 2001; 2002b), but obviously he does not see it in that way.

So why this disconnect? The opposite claim that *all* maps are political has a long
pedigree. One of the first to powerfully grasp this point was Denis Wood in his
book *The Power of Maps* in 1992 (from which the "Marble-hearted" quotation comes).[2]
Wood's book was a kind of extended "catalog" for a successful exhibition of the
same name that he had mounted at the Smithsonian museum (Figure 2.2), and as

his ultimate test case Wood undertook to show that even something as "neutral" as a topographic or road map could have a politics (or "interests" as he put it, which is not exactly the same thing, but close enough for our purposes). His analysis of the North Carolina state road map is something of a tour de force and is now a classic of cartographic writing. Wood showed – in a similar way to Harley (see Chapter 7) – that the silences and omissions of the map (for example any non-automobile paths or bikeways, or any information about public transportation) represented a deliberate privileging of the car and thus a deliberate (and political) choice. The map was a series of tubes for cars. (A similar debate has arisen more recently in the context of Google Maps, which from the beginning offered a "drive to" option for getting driving directions; they have now added public transport and "walk to" options in some cities.)

Wood went on to analyze the "marginal" information on the map, which he rated as just as important as the map information. Drawing from the work of the French cultural critic Roland Barthes, who found mythologies in the everyday things we surround ourselves with, Wood deconstructed the image of the governor and his family (leaning against a car, naturally!), and the underlying anti-union, pro-business message put forth by the Chamber of Commerce. Wood's argument therefore is that maps are bound up within (and may even help create) a whole regime of claims, truths, and ways of seeing. More recently he has suggested the words "perimap" to refer to those discourses that "surround" a map (Wood and Fels 2009).

But what does it mean to say that maps are political? Surely in one sense that's quite plain now. Topographic maps for example are made by the state. Is that what we mean – that this map is political because it is made by a branch (or branches) of government? That is quite a constrained view. And as we will see in the great Peters Map Controversy, many cartographers and GIS users simply don't agree that mapping is or *should be* political, and they equate politics with ideology or even propaganda. Perhaps more significantly, there are also plenty of people who find any reduction of mapping and GIS to questions of "power" and politics to be if not irrelevant then only part of the story. A good example would be Denis Cosgrove, who sought to interpret the map as a cultural artifact that "excites imagination and graphs desire" (Cosgrove 1999: 15) – a "poetics of space" (1999: 17). We shall come back to this interpretation in Chapter 12.

In order to avoid the usual pitfalls of argument about whether maps are (or should be interpreted as) political, I will suggest that it is better to understand them as part of the *political economy of government*.

The Political Economy of Government

The phrase "political economy" has deep historical roots and is associated with the emergence of the modern political state in the eighteenth century. Rousseau's entry in one of the key texts of the Enlightenment, Diderot and D'Alembert's

Encyclopédie, in 1755 is well known. The question for Rousseau and his contemporaries was how this emerging modern state should be organized, given that the ancient method of rule by dictatorial sovereignty no longer held up. The Machiavellian absolute rule recommended to "the Prince" (the sovereign ruler of a state) was no longer possible nor desired (for one thing it was terribly expensive). Put simply, Machiavelli urged that the prince look to his territorial boundaries and defend them from enemies foreign and domestic. Rousseau began his discourse on political economy by pointing out that the word economy had its origins in two Greek words: *oikos* or hearth and *nomos* or law. Economy then, was originally the "wise and legitimate government" of the household for the common good. This idea could be and should be (argued Rousseau) applied to the government of the state.

What did this mean in practice? Well, it meant the "right disposition of things," a sort of rallying cry around which politics could be organized.[3] Where should things go? How many things were needed? Which parts of the country were doing well and which were not? What in fact did it mean to be "well ordered"? It was this cry or idea that Michel Foucault discusses in his lectures on governmentality:

> The things government must [now] be concerned about . . . are men in their relation-
> ships, bonds, and complex involvements with things like wealth, resources, means of
> subsistence, and of course, the territory with its borders, qualities, climate, dryness,
> fertility, and so on. (Foucault 2007: 96)

Governmentality is that rationality (thought, practice, conceptualizations, and discourse) that asks the question, What is government and how is it practiced? The "art of government" is the question of who can govern, what is governed, how best to govern, and what governing is (Gordon 1991: 3). It is the *political* question of governing. It thus includes efforts to administer hospitals, prisons, schools, businesses, the family, the police, and even the self. This was a more expansive notion of territory not just as a bounded unit which needs defending (*à la* Machiavelli), but one in which there is more of a notion of environment or "milieu." It is also quite a bit different from saying that the state is the only or prime institution involved in government.

Foucault pointed out that political economy represented an important shift in the object of analysis. As he had discussed at length in previous work, especially his most well-known book, *Discipline and Punish* (Foucault 1977), the focus of power control through "discipline" was often targeted at the individual. Thus his discussion of the well-known panopticon of the social reformer and architect Jeremy Bentham, with the prisoners arrayed in their cells and observed from a central vantage point.[4] But with political economy the focus is not in individuals but on *populations*. Foucault says:

> After the anatomo-politics of the human body established in the course of the
> eighteenth century, the emergence of . . . what I would call a "biopolitics" of the human
> race. (Foucault 2003b: 243)

In actuality, with the advent of governmental rationality in seventeenth- and eighteenth-century Europe, there was an increased shift to administering resources constituted primarily as a population, and to manage those resources for the best possible ends. Residents of a territory were no longer subject an all-powerful sovereign who could freely expend their lives, either in defense of his own life or for defense of his rule when externally attacked (i.e., through warfare). In the modern state, the concern for life is to administer it through controls and regulations so that resources might be rightly apportioned. Foucault traced two specific aspects of this shift: one centered on the individual through discipline and optimization of capabilities (*anatomo-politics*) and one centered on the population as a "species body" or a *bio-politics* (Foucault 1978: 139, see also 25–6). This shift had occurred by the eighteenth century, a sequence that is reflected in Foucault's own work by a shift from *Discipline and Punish* (Foucault 1977) to governmentality.

Political economy (as it developed out of the theories of Rousseau but also Adam Smith and David Ricardo) became a legitimate part of government in the sense that it provided knowledge of the state's activities that were not just the imposition willy-nilly of an all-powerful sovereign.

There's quite an interesting analysis about just what political goals the state set following the demise of sovereignty and the emergence of the modern political state, but the point to bear in mind here is that of knowledge. Because it needed knowledge of its territories, and its peoples (both individually but increasingly as groups or populations, although it was not immediately apparent just what constituted a coherent population), we can now understand how mapping came to be such an important part of the political economy. This concept of the milieu has played an extremely important role in understanding the relationship between the environment and the distribution of things, especially for this emerging idea of populations, and it is here that mapping is important. If you're going to understand populations and their needs and so on, it quickly becomes apparent that there are differences between different groups. Since the days of exploration in the sixteenth century Europeans had encountered a bewildering array of different peoples in Africa, Asia, and America.

How to understand these groups? There were two main answers to this. In the 1740 edition of his work *Systema Naturae* the great classifier Linnaeus posited four geographical subdivisions of humans: white Europeans (who were sanguine, pale, muscular, and clever), red Americans (who were combative and choleric), yellow Asians (who were melancholic, severe, and avaricious), and black Africans (who were slow, relaxed, and negligent – all these descriptors are Linnaeus's own). For Linnaeus these were natural categories (although they didn't work out very well geographically due to exceptions and overlaps). What's more these populations were fixed since the time of their original formation (Linnaeus rejected theories of evolution). For Linnaeus and those who have since followed him, the way to understand populations or species was through their similarities. What features did they share? What was a member and sub-member of a particular class? How did they relate?

Classification is obviously an extremely important part of the way we order the world (Bowker and Star 1999). By grouping many things into a smaller number of things we take advantage of a coping mechanism. Linnaeus applied his system with gusto, not only to animals but to plants and "minerals," and it has remained an important organizing principle (class–order–genus–species) within scientific taxonomy for 250 years. If, in mapping, you want to determine who is where, which populations occupy which territory, or how the human races are distributed, then the classification approach can provide an answer. So when, for example, we want to look at the results of the census, whether from a hundred years ago (Walker 1874) or ten years ago (Brewer and Suchan 2001), you can work with population groups founded on their members sharing common elements.

The anthropologist Jonathan Marks says that it was this style of thinking that dominated until about 40 years ago in anthropology (Marks 1995). It was particularly influential in dominating people's understanding of race (and this goes beyond anthropology to include other disciplines such as sociology and geography), that here were these groups based on common elements that they shared that could account for the distribution of people across the globe. And for Marks this classificatory system, which would somehow keep finding the same small number of Linnaean racial groupings (although the exact number varied), was incredibly damaging to our understanding of diversity. Marks notes that in his own lifetime Linnaeus was opposed by another thinker, the Count de Buffon, who was the leader of a sort of opposition to Linnaeus. While Buffon also opposed theories of evolution and believed that species remained stable since their formation, he did accept that *within* species the environment (milieu) could cause distinct population differences. So where Linnaeus had classification as his goal (isolation of common elements), Buffon had *diversity* (explaining variation). Buffon asserted that the environment and especially climatic temperature operated to produce changes (he suggested taking an African population to Denmark to see how long it took them to turn white). So with the rejection of Linnaean classifications, the focus in studying populations can now turn to understanding the range of diversity in human populations, without the over-arching categories of race. How do populations adapt to environments? What is the role of genetic drift? These are nowadays the main questions in explaining human diversity. The milieu shapes not races but human population groups.

So these two ways of approaching populations frame our view of how mapping was part of the political economy.

From *l'état, c'est moi* to *l'état, c'est l'état*

When King Louis XIV of France came to power in 1661, French cartography was a distinctly second-class affair. Both England and the Netherlands far exceeded France in terms of mapmaking, surveying, navigation – even in making instruments for measurements. Only a few years later however France could claim superiority over the fading efforts of its rivals. The French produced a renowned national survey under the direction of the Cassini family over the course of some 100 years. The patron of this survey was none other than the king's minister, Jean-Baptiste Colbert, who had examined France's maps for both quality and quantity and found them in a deficient state. The story of how this remarkable turnaround was achieved bears witness to the role that maps play in political economy. Colbert's explicit goal was to restore France's interests and to strengthen the authority of the state. In the course of these developments mapping entered several new dimensions; the accurate mapping for the first time of state political boundaries, and the development of "cartes figuratives" – thematic maps (Konvitz 1987). Both of these mark an important shift that began in the seventeenth century, deepened in the eighteenth, and matured in the nineteenth. All along the way, maps and political economy were interrelated.

A recent article by Christine Petto neatly captures the shift from sovereignty to political economy in eighteenth-century France by examining two influential cartographers, Alexis-Hubert Jaillot and Guillaume Delisle (Petto 2005). For Jaillot, a commercial map publisher who worked in the second half of the seventeenth century, mapping was part and parcel of the glorification of the sovereign, King Louis XIV (the "sun king"). Louis was a great patron of the arts and sciences, and his Versailles palace was filled with writers, artists, and scientists, of whom Jaillot was one. The explicit purpose of a lot of the works produced was to burnish the kingly image: *l'état, c'est moi*:

> Whether in painting, sculpture, inscription or tapestry, the presentation of the king followed a "rhetoric of the image" developed in the Renaissance. The body position, the regalia, the mythical portrayal, and the total subject matter were all part of this rhetoric and helped form a "cult of image", in this case the image of Louis XIV and the cult of the Sun King. (Petto 2005: 55)

Maps played an important role in this equation; their margins would show the king standing over a defeated soldier (the *frondeurs* or insurrectionists who defended rights and liberties against royal encroachment during the Thirty Years War), or dominating a three-headed Cerberus, meant to represent the threat from the Triple Alliance of England, Spain, and the United Provinces. Petto notes that "the alignment of power and mapping not only involved the use of these motifs in the design elements of the map but also reflected the overwhelming power of Louis's personal state" (Petto 2005: 55). This was clearly a matter of maps in the service of the sovereign.

Jaillot's participation in this patronage eventually resulted in him being made Geographer to the King in 1686 as well as receiving a generous annual salary. By the time he died in 1712 however, the identification of the state with the person of the royal sovereign was weakening. A new form of government cartography began to emerge, grounded in scientific authority and the delineation of territory – what Petto calls *l'état, c'est l'état*. Eighteenth-century France mapmakers were increasingly able to take advantage of new and more accurate territorial surveys, such as those done by the Cassini family using powerful new telescopes and astronomical observations. (Louis XIV reputedly complained that these surveys, although accurate, shrank his kingdom by 20 percent)

Guillaume Delisle was active a little later than Jaillot, from the early eighteenth century. While he certainly published maps for the king (Louis XIV after all didn't die until 1715), there was a shift of emphasis. No longer designed to promote the person of the sovereign, Delisle's maps were produced for the state as such. Delisle's particular interest were foreign maps, especially those of the Americas. Much of the information about these places came from traveler's reports or sketch maps and it was not always apparent how reliable this information was. The assumption was that only astronomical surveys would be reliable and this meant an increasing emphasis on scientific approaches. France's foreign mapping interests paralleled its increasing commercial and mercantile overseas interests in sugar, slaves, and fur. Delisle was heavily involved in this type of mapping (for example of the Louisiana territories in 1718 which set off a kind of "map war" by cheekily extending French boundaries into territory claimed by the British, which was rebutted by a series of English maps).

It is an over-simplification to say that mapping at this date and place shifted from lauding the sovereign to serving the state using scientific methods. Delisle sought patronage as eagerly as Jaillot, and received it when he became "*First* Geographer to the King," a title which eclipsed that of Jaillot. Nor was he the first to take advantage of astronomical surveying – Cassini moved to France as early as 1669, where he set up the Paris Observatory, and French cartographers had tried to fix territorial boundaries of their country since the 1560s (Buisseret 1984), although this was at the time more of a legal procedure than a scientific one.[5] Nevertheless the lives and approaches of these two men do mark out the shift in attitude of mapping as a form of governmental knowledge that would only intensify through the eighteenth and especially nineteenth centuries. The kinds of questions to which Delisle addressed himself, namely the truth of the territorial knowledges that were being produced, were quite different from the ones concerned with promoting sovereignty. The question of truth was put into play; the question of what was known and what was the best way to know, and what was the relation of truth to knowledge – these were the new questions in which mapping was now taking part.

Thematic Maps and Governmentality: Introducing the Calculating Surveillant State

So far we've seen that the modern state used maps in its assessment of populations and their distributions across territories, in other words as part of the political economy. But the way the state did this wasn't uniform.

Historically there were significant changes in the way that population was understood. In the early nineteenth century when thematic mapping was invented by political economists, the question of population was one of knowledge about particular places as defined by political boundaries, and the people who occupied those spatial units. Places were assumed to exist, and the question was: what were they like?

However, by the early twentieth century, when cartography was first developed as a discipline by cartographers, the emphasis got reversed: now the question was: how did the density and characteristics of the population *make* places? This is the way we still understand place today; as placemaking and the production of space. This is perhaps a more useful way as it allows for the recognition of contested place and counter-productions.

Many GIS projects in local government, for example, are directed to mapping resources such as traffic flows, and this has implications for emergency vehicle routes, traffic-light timing patterns, speed limits, deployment of police, and so on. Federal government wishes to manage its parklands and wildlife areas (the first GIS in the 1960s was a government inventory of Canada to identify land resources and their use). Businesses wish to know where their potential customers are. All have in common the idea that resources need to be managed and risks to depletion of those resources avoided. Governmentality brings to the fore the fact that data collection and analysis are not done in isolation from specific governmental goals and ends as a political question and in a particular way.

There is no one particular moment when this reversal occurred, a before and after. In some ways it never happened. But I want to pick out two historical moments that exemplify what was going on. The first, and drawing on the detailed work of Matthew Hannah, is the 1874 *Statistical Atlas* of the first census following the Civil War (Hannah 2000; Walker 1874). The second, 40 years later, are the boundary-makers of World War I and the way they redrew the political map of Europe based on race (Crampton 2007b; Winlow 2009).

Despite that fact that mapping has been around for thousands of years, thematic mapping is a totally modern form of mapping and was developed and refined only from the late eighteenth century onwards. By the 1850s most of the types of thematic maps in use in today's GIS had been invented. The choropleth map, for instance, was first used in 1826 (Delamarre 1909; Robinson 1982). Choropleth maps were a brilliant example of how thematic maps "speak to the eyes," as William Playfair wrote in 1802, and thus they contributed to a discourse of resource assessment. Playfair's own groundbreaking work on graphical statistics was itself directed toward political ends. His graphs of the balance of trade, for example, were published to urge the government to reduce its debt obligations to other countries (such as America).

The 1826 choropleth map depicts the ratio of (male) children in school to the population of each department in France. It was used by its author, Baron Charles Dupin, to identify what he called *la France obscure* and *la France éclairée*, the unenlightened and enlightened regions of the country (Dupin 1827). Scarce resources could now be targeted more efficiently (such as the building of more schools). There was tremendous contemporary interest in this map, and it was later credited with increasing the number of schools in France. It was much copied, and almost identical maps on a range of subjects (crime, education in the Low Countries, etc.) soon appeared. Yet it is significant that Dupin was not a cartographer (he was a mathematician and economist, and later had a career in politics); he had no interest

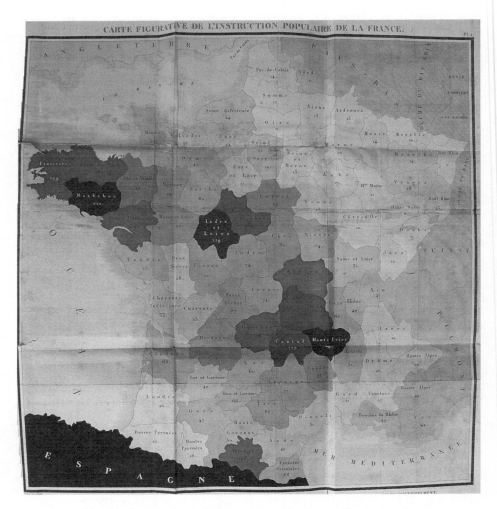

Figure 6.1   The first known choropleth map, 1826, by Baron Charles Dupin.
Source: photograph from Dupin 1827.

in the map as a map (or the fact that he had invented such an important map type); rather, he was passionately concerned with political economy and the politics of production, distribution, and consumption of resources. Dupin made a direct linkage between the health of the nation and the education of the population: what is good for the nation is good for the population and vice versa. The population is a resource for the country (Figure 6.1).

The 1874 census atlas

This rationality was late in starting in America in terms of statistical mapping, but it was finally spurred into action by the Civil War. The first census following the war took place in 1870, when there was an overwhelming need to take stock of the country. Matthew Hannah's analysis (Hannah 2000) of the 1870 census can therefore be seen as a study of the middle passage between the initial invention of thematic maps and the later "reinvention" of thematic mapping in the early twentieth century in terms of how place was "calculated." The calculation of space is a term that some scholars have used to investigate the particular governmental production of space as a knowable sets of resources, often at risk (Elden 2007; Hannah 2009). I will return to it again in the context of race in Chapter 11.

Because the Walker *Atlas* is the first statistical atlas of the census in America, it brings together developments in several areas that were necessary for mapping to flourish. First, there was a desire and need for the nascent modern state to know about itself and its constitution through the prosecution of national censuses. Second, the invention of statistics and probability in the first half of the nineteenth century allowed for calculation and measurement. But what was to be measured and calculated? It was, third, the population distributed over the territory. Therefore we have a triad of political cartography: government, statistics, and population, or Foucault's notion of biopower. The *Atlas* is a hinge that joins the political economists and their choropleth maps with the cartographers of the early twentieth century (see Figure 6.2).

A rash of publications in the early 1900s illustrates this concern with populations, most notably by Mark Jefferson (1863–1949), a cartographer and geographer at

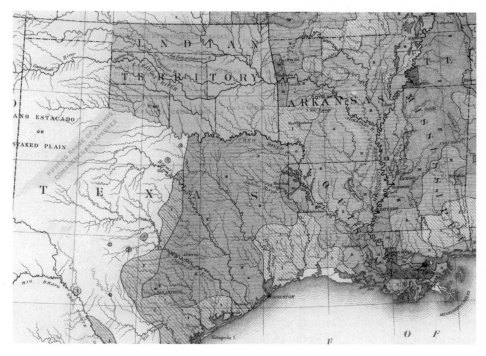

Figure 6.2 Density of the "Constitutional Population" in Walker's *Atlas* (detail).
Source: Photo by author, Plate 19 (Walker 1874).

Eastern Michigan University (Martin 1968). Jefferson's population work benefited from his innovative cartographic skill and experience (he was the Americans' chief cartographer at the Paris Peace Conference in 1919), to give him a sense of where people lived and why.

Jefferson was more interested in the meaning of the place in which people lived, rather than in defining places by political borders. In this sense Jefferson was markedly ahead of his time, in that he understood place as actually experienced space, an understanding that would later be influentially articulated in political geography (Agnew 2002). We see here, then, the explicit desire to understand place as it was constituted through people and their actual lives, rather than the approach of the nineteenth-century political economists, who wanted to know what kind of people lived in certain politically predefined areas.

"Thinking out space": The Americans in Paris

In 1917 the United States entered the war and simultaneously began preparing for peace by establishing a secret research group called the Inquiry. Instituted by President Wilson and headquartered at Isaiah's Bowman's American Geographical Society (AGS), the Inquiry was charged with determining American policy to be used at the presumptive peace conference (held in Paris from January–June 1919). To redraw the map of post-war Europe the Inquiry sought to isolate both *identity* and *territory*. The peoples or population inside bounded segments of space (regions) should be all alike in the crucial respects. While language had partly been a guide to this since the nineteenth century (Dominian 1917), the ultimate goal was *racial partitioning*. If these territorial units could be identified then this would lead to stable sovereign states across Europe who would be unable to claim extra territories on the basis of racial affiliation of occupants. In other words, not only could distinct natural races be identified, but if their areal extent could be unambiguously determined this would yield viable and peaceful sovereign states.

The Americans in Paris had a particular way in which they "thought out space," as Foucault once described this kind of planning (Foucault 1984: 244); they used race and racial boundaries. The following document sets out their plan (Figure 6.3).

Clearly the Inquiry had no wish to impose arbitrary lines across Europe but rather ones which were scientifically justified. President Wilson made vague pronouncements that America would bring about "the will of the people through self-determination," but who *were* the people and *where* were they? The members of the Inquiry believed that if they could draw careful lines around each race or at least nationality then it would leave everybody happy and prevent further war. Today we are familiar with this idea from the Israel/Palestine and US/Mexican borders among others. But for Wilson's researchers this was a new question. In the old days, the colonial diplomats placed their lines without too much regard for populations and aimed instead for strategic defensible borders to keep "the natives in their places" (Noyes 1994). Good boundaries therefore ran along mountain tops, rivers, or other topographical features (Holdich 1916). One of the most influential people who held this view was

III.—How Inquiry can help with regard to each major task of Conference

1) *Boundaries:*
   a) Racial boundaries:
      i) Make a racial map of Europe, Asiatic Turkey, etc., show-ing boundaries and mixed and doubtful zones.
      ii) On basis of *i*) draw racial boundary lines where possible, i.e. when authorities agree; when they disagree select those we had best follow; when these disagree map the zone of their disagreement; study density and distribu-tion of peoples in these zones.
      iii) Study, in each case, the stability or instability of racial distribution (e.g. Macedonia, N. E. Albania) as af-fected by change of political boundaries and conse-quent governmental action, by economic forces, by religious forces, by other cultural forces, etc., but all with stability or instability in mind.
   b) Historic facts and national or racial aspirations as indicating boundaries (e.g. Serbo-Bulgarian '12 agreement).
   c) Economic facts and needs as indicating boundaries (e.g. Jugo-Slavia or Albania or Poland or Czecho-Slovakia as a well-balanced economic unit, access to ports, and markets, i.e. minor units that should not be disrupted, etc.).
   d) Defensive needs as indicating boundaries.
   e) International commitments and obligations as affecting pro-posed boundaries.
2) *Government:*
   a) Inquiry can give some account of political and economic and military strength and weakness of "states," and of what participation in government peoples have had, and an estimate of their capacity for self-government.

Figure 6.3 Inquiry Document 893 "A Preliminary Survey" (no date but probably late July 1918). Source: FRUS (1942–7: Vol. I, 20).

Halford Mackinder. In 1915 Mackinder warned that there would be little "ideal map-making" of the sort the Inquiry was busy planning for, but rather it was a case of realpolitik and clipping Germany's wings if the allies were to "conquer that power" (Wilkinson et al. 1915: 142). This realpolitik was familiar, for just ten years previously Mackinder had published his famous "pivot" or heartland map, docu-menting the transition from sea- to land-based power (Figure 6.4). O'Tuathail has called this 1904 map "perhaps the most famous map in the geopolitical tradition" (O'Tuathail 1996: 31).

The debate between those who thought as President Wilson and the Inquiry did, that frontiers should go around homogeneous populations, and those who saw themselves as realists, is an important one. If Mackinder's world view informed the Cold War (Dodds and Sidaway 2004) and a century of geopolitics, population-based mapping has been instrumental in the neoliberal calculating surveillant state. In order to know about these populations the state must deploy a panoply of surveillance methods, of which the census is only the most obvious. Where some

THE NATURAL SEATS OF POWER.

Pivot area—wholly continental.   Outer crescent—wholly oceanic.   Inner crescent—partly continental, partly oceanic.

Darbishire & Stanford. Ltd.   The Oxford Geog! Institute.

Figure 6.4   Mackinder's pivot or heartland map. Source: Mackinder (1904).

points of its territory are seen as more risky, such as borders (Amoore 2006) then extra surveillance will have to be performed.

The effort put into the question of Europe's borders and the level of detail the Inquiry was prepared to create was enormous.[6] Literally thousands of documents, reports, maps, and statistics were painstakingly researched, prepared, written, and approved by the 150-strong team. Practically all of the well-connected or well-known academics in geography, history, economics, English, and even medieval studies were hired. These included Isaiah Bowman, Ellen Churchill Semple, Charles Seymour (later President of Yale), labor writer James Shotwell, the anthropologist and eugenicist Charles Davenport,[7] Armin Lobeck, the physical geographer Douglas Johnson, and many others.

Members of the Inquiry undertook lengthy fieldwork trips, including the well-known journalist Walter Lippmann who went to Europe to interview the many representatives of countries with disputed territory and leading men of the time (including H. G. Wells). Mark Jefferson, Isaiah Bowman's old professor, was hired as Chief Cartographer and most of the third floor of the American Geographical Society was devoted to the Inquiry, guarded by passwords and a night watchman.

In the following two previously unpublished maps note the incredible care and attention to detail that the Inquiry took to visualize the terrain and to place the territorial boundaries. In the case of disputed international borders the Inquiry drew on ethnographic surveys to produce the "linguistic line," that is, the ideal of population division according to which language people on the ground actually spoke (Figure 6.5). It shows that the "line claimed by Italy" intruded well into non-Italian speaking territory. In Lobeck's hand-drawn map of the area almost every sinkhole in the limestone around the disputed city of Fiume is easily visible (Figures 6.6 and 6.7). Bowman called Lobeck one of "only four men in the country" who could draw these maps. Obviously this was more than just "have a go mapping." It would have to stand up to some strict scrutiny; not just legal and political, but in the court of world opinion as well (the Paris Peace Conference took place in a blaze of publicity and the President of the USA and his staff took over an entire hotel for six months). Lobeck's maps were not puny affairs but five feet wide and broad, with a vertical exaggeration of four times to show the detail.

Thus the question for the Inquiry was simultaneously one of *knowledge*, especially territorial and spatial knowledge (very specialized knowledge in the case of Lobeck), and of a *rationality* or reasoned basis on which to deploy that knowledge based on race.

Summary

We have seen how the Inquiry took as its predicate the derivation of scientific racial boundaries across Europe; boundaries that were assumed to reflect an underlying racial partitioning that could be discerned on maps. The Inquiry knew well that

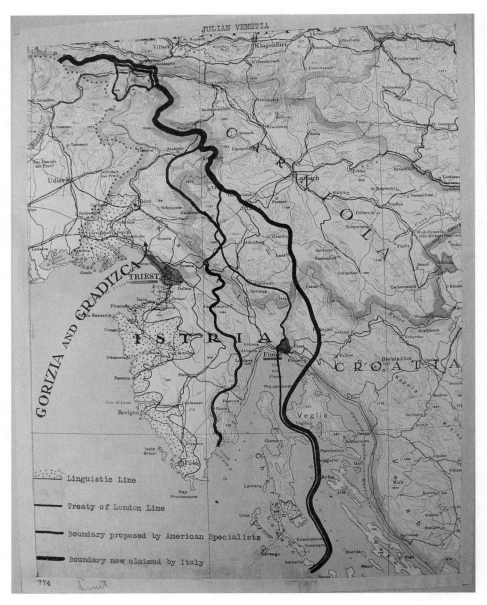

Figure 6.5   Disputed international borders; the "linguistic line," the line proposed by the Inquiry, and the line proposed by Italy. Source: NARA RG256, Entry 55 Italy.

this was not a simple reduction of identity to space. But they assumed that territory and its rightful populations could be discerned if you looked hard enough and assembled the right data. Once "propaganda" ("politics") had been removed from the equation, a clear track could be cut through the morass of competing claims. As the Serbian geographer Jovan Cvijić put it in the pages of the *Geographical Review*,

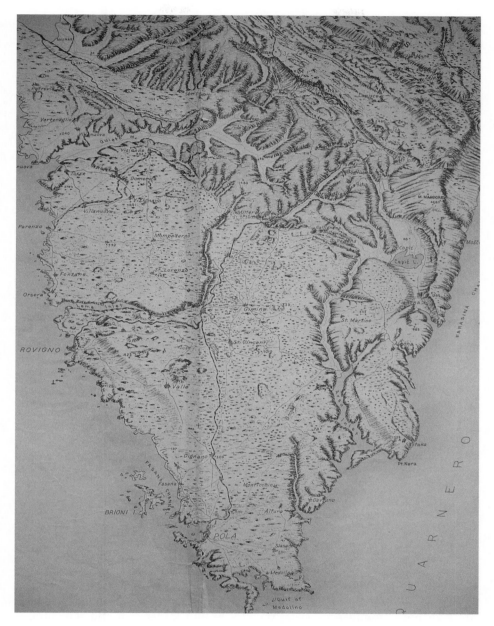

Figure 6.6   Part of Armin Lobeck's map of the Istrian Peninsula. Specially prepared for the Inquiry, this map shows the level of care and detail taken. Source: NARA RG256, Entry 45.

his ethnic fieldwork allowed him to discern "natural barriers" in Europe which picked out "zones of civilization" (Cvijić 1918: 470). Yet anthropology has taught us the opposite. Speaking of racial divisions the anthropologist Jonathan Marks observes that "[w]e don't know how many there are, where to draw the boundaries between

Figure 6.7   Detail of Armin Lobeck block diagram map. Source: NARA RG256, Entry 45.

them, or what those boundaries and the people or places they enclose would represent" (Marks 1995: 275). (See Chapter 11 for more on race.)

So the bottom line is that one useful way to understand mapping and politics is as technologies of government. Not just the state level either. There are all sorts of levels of government beyond "the" government, including government of the city, neighborhood, family, or even government (technologies) of the self, a topic Foucault takes up in several later works (Foucault et al. 1988).

Finally, let's consider an objection to a governmental approach to political economy: does it do without a theory of the state? In the context we can consider Scott's influential book *Seeing Like a State* (Scott 1998), and his concept of legibility, that is, the way the state sees and deals with its "right disposition of things" (through house-numbering, cadastral surveys and grid plans, and standardization of knowledges). Scott argues that centrally planned, map-driven states are never as successful as those which incorporate local knowledges. In a recent paper Hannah identifies two key components of what he calls calculable territory; this *legibility* and the ways that knowledge of territories (including through GIS, GPS, topographic maps, and so on) is *mobilized* (Hannah 2009). If the state is busy calculating territory, how can we do without a theory of the state (and perhaps its failures)?

In *Birth of Biopolitics*, Foucault poses this question to himself, answering that "Well, I would reply, yes, I . . . must do without a theory of the state, as one can and must forgo an indigestible meal" (Foucault 2008: 76–7). All very well, but this is hardly likely to convince those who track post-9/11 encroachment of the state on civil liberties and the advance of the surveillant state. More seriously therefore:

> What does doing without a theory of the state mean? If you say that in my analyses I cancel the presence and the effect of state mechanisms, then I would reply: Wrong, you are mistaken or want to deceive yourself, for to tell the truth I do exactly the opposite of this. Whether in the case of madness, of the constitution of that category, that quasi-natural object, mental illness, or of the organization of a clinical medicine, or of the integration of disciplinary mechanisms and technologies within the penal system, what was involved in each case was always the identification of the gradual, piecemeal, but continuous takeover by the state of a number of practices, ways of doing things, and, if you like, governmentalities. The problem of bringing under state control, of "statification" (*étatisation*) is at the heart of the questions I have tried to address. (Foucault 2008: 77)

Perhaps one way to think about this then is that governmentality works hierarchically across a range of untidy geographies, from the individual to the territorial to the state, and that it has converged with state-theorists "if only partially, hesitantly and with recurring expressions of lingering mistrust" (Hannah 2009: 66). Jessop, concurring (2007), finds that governmentality is not self-evidently incompatible with Marxist theory:

> it seems that, while Marx seeks to explain the why of capital accumulation and state power, Foucault's analyses of disciplinarity and governmentality try to explain the how of economic exploitation and political domination . . . this re-reading shows that there is more scope than many believe for dialogue between critical Marxist and Foucauldian analyses. (Jessop 2007: 40)

Hannah's argument in other words is that the two threads of the argument (governmentality and Marxist-inspired state theory) come together over the concept of the calculable territory. If this chapter has dwelt on governmentality it is not because this claim is incorrect (it still needs to be studied), but rather that it is more cartographically tractable to examine the relationship of the state to mapping. My discussion here is suggestive rather than definitive, and we'll continue to address it in the next few chapters!

Notes

1   Wood is quoting from the unpublished notes of a lecture given at Penn State in 1991 by Brian Harley, parts of which later appeared as Harley (1991). Harley died just a few months later.

2   See *King Lear*:

> "Ingratitude, thou marble-hearted fiend,
> More hideous, when thou show'st thee in a child,
> Than the sea-monster."
> *King Lear*, Act 1, Scene 4, Line 283

3   This phrase was noted by Foucault in his Lecture at the Collège de France on February 1, 1978, where he contrasted the Machiavellian approach to a later one emphasizing government, La Perrière's *Le Miroire politique* (1555). This lecture appears in the recent "official" translation (Foucault 2007) but has been available in samizdat form for nearly two decades (Burchell et al. 1991).

4   Contrary to popular belief that these ideas were purely theoretical, hundreds of prisons around the world were built modeled on this design, not to mention innumerable school classrooms, asylums, and other institutions. I discuss one them, the Eastern State Penitentiary in Crampton (2007a).

5   Buisseret notes that legal representatives would gather at the disputed points of the frontier where the village's oldest residents would be interviewed and deposed (Buisseret 1984).

6   Yet surprisingly little is still known about the Inquiry (much less, for example, than is known about Mackinder). With few exceptions its geographical, political, and cartographical work has gone unexamined. Yet the Americans were arguably the most influential voice at the Paris Peace Conference. Created by President Wilson's "national security advisor" (as we might call him today), Edward M. House, it gave rise to the Council on Foreign Relations and in the UK the Royal Institute of International Affairs (Chatham House).

7   Davenport, who was America's leading eugenicist at the time, was hired by the Inquiry for the spring and summer of 1918 to prepare reports on ethnic populations in problem areas. For more on Davenport see Crampton (2007b).

Chapter 7

The Political History of Cartography Deconstructed: Harley, Gall, and Peters

It's a perfect late spring day in May 1977, temperatures are hovering around 70°F, and two men, one a local and one a visitor from London, are strolling along a footpath toward All Saints' Church in the village of Highweek just to the north of Newton Abbot in Devon. Bella, the family's boxer dog, is rooting around in the verge and occasionally disappearing off into the fields, but neither man is worried about her getting lost for it is a very familiar path. "The usual route . . . [was] east along Knowles Hill Road, north through the meadow near the right edge of the map, to work our way over to the pathway leading up to Highweek Church, and to circle back home through the residential areas" (Woodward 2004).

Neither man had difficulty with the mild slope up to the church, and in any case at 45 the dog's owner could still benefit from his youthful pastime of cross-country running. The hill is not that steep and the path unwinds gently along the side of it. The small town off to their left is lost between the trees and their destination, a church and its World War I war memorial, is not even at the top.

The journey to Newton Abbot was a pleasurable three-hour journey from Paddington Station in London, and both men were glad to get out of the freezing office in Brian's home at number 6, Knowles Hill Road (the space heater was strictly "on demand" only). About half way along the church path "where the hedgerows opened up to a field on the right" (or as we might also say more precisely, but not necessarily more meaningfully, at 50° 32′ 15.4″ N., 3° 37′ 05.8″ W.), their discussion took a fateful turn. The men, both academics, were discussing history and specifically the history of mapping. The two men (but this is awkward, their names as you've guessed are David Woodward and Brian Harley) have long agreed that the current state of affairs in this subject is entirely unsatisfactory. Brian, at least, was prepared to write four volumes that would cover North American mapping, with its tremendous history and driving narrative during the age of exploration through to the close of the Victorian era. He was coolly relaxed about the idea that this would be a "fifteen year project – probably for the rest of my active life!"

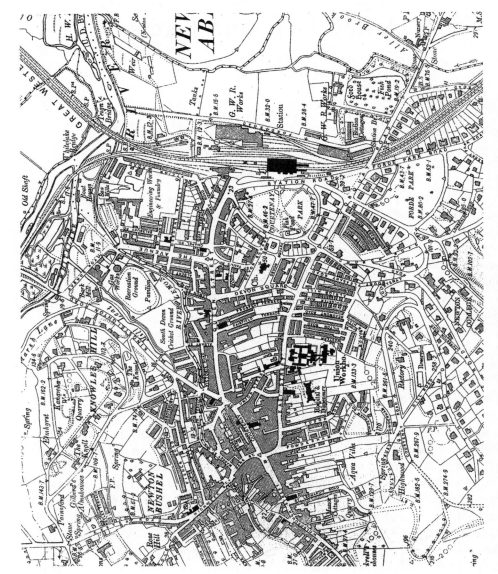

Figure 7.1 Newton Abbot on the 6-inch Ordnance Survey map. Source: Harley (1987).

David countered with the suggestion "why not a four-volume *general* History of Cartography?" The idea seemed a natural; they could both edit them (taking say, 250,000 words per volume for a total of 1 million words), and as David said "it would take no more than *ten* years of our lives!"

In fact, ten years later only volume one of a now projected *six*-volume series had appeared (Harley and Woodward 1987), and as I write this in 2008 publication has been made only through volume three (Woodward 2007). In the remaining 14 years of his life following "the Devon walk" the church toward which they were walking that day was freighted with memory for Harley; not only did his daughter marry there, but "it is also the place of sadness. The ashes of my wife and son lie buried against a north wall of that churchyard" (Harley 1987: 20), and he himself would be buried there within a couple of years of writing those words. (Harley's wife and son had died separately in 1983.)

Brian Harley made so much impact on historical geography, the history of cartography and mapping in general that in this chapter we shall briefly examine what he had in mind that May morning. As one speaker at Harley's memorial noted "Brian Harley did more for the history of cartography in the past ten years than everyone put together had done in the previous fifty" (G. M. Lewis 1992: 19).

Those are pretty strong words, so let's see what Harley and Woodward thought they were up to:

Figure 7.2 J. Brian Harley in 1988. Photo courtesy of Ed Dahl. Used with permission.

> From our experience with prehistoric, classical, and medieval cartography, it rapidly became clear that maps could be seen from very different angles, not just as measured, objective representations. Harley argued that the multi-volume History had to be more than a reference book; it had to blaze the trail for new interpretive approaches to the material-approaches that would tease out the meanings in maps that were not apparent on their surface. It should broaden the scope of the volumes beyond maps representing the physical milieu to the maps of imagined cosmographies wherever these delineated structured concepts of space. This broadening forged new links with scholars in other disciplines – such as history, art history and literary criticism – which have been enormously rewarding. (Woodward 1992b: 121–2)

Blazing a trail, forging new links. Not an easy task and one that stirred up opposition as Woodward admitted in an interview in 1995:

> I had to personally redefine the idea of what a map is for myself. . . . The Western idea of a map is limited. We think of it as systematic, geometric, that has scale and projection, that you can measure off of and that people think is absolutely true and unbiased. . . . [Asians by contrast] were much more interested in capturing the spirit of the land . . . we've had criticism from geographers saying "these are not maps; these are religious statements" . . . personally I'm glad we've done it. It opens links to religion, literature and culture. (Hayes 1995)

Harley is best known today for his paper on deconstructing the map (Harley 1989a), but the thing to bear in mind about him is that he was not a theoretician. On the contrary, his most respected work includes some very finely ground empirical accounts of Ordnance Survey maps and production techniques. Nor was he an ideologue; the key to Harley is that he was a great one for trying things on. There is no key theoretical position that I can offer you that would sum him up. He tried a variety of approaches including iconography, semiotics, art history, a taste of French deconstruction – all of it could be picked up, used, half thought through and then dropped when the next shiny object came into view.

Except there is this which could possibly be a motto or manifesto (not a theory!):

> I believe a major roadblock to understanding is that we still accept uncritically the broad consensus, with relatively few dissenting voices, of what *cartographers* tell us maps are supposed to be. (Harley 1989a: 1, emphasis in original).

When his friend and colleague Matthew Edney wrote his obituary it had the word "questioning" in the title three times (Edney 1992). Don't believe the hype! It's this that marks Harley as a critical cartographer more than anything else (see Chapter 2 for more on the critical attitude). If he's known mostly outside cartography as a theory guy, he wrote at a time when theory in cartography was a pretty arid place. He himself was not much of a theoretician – his mind was far too empirical for that. What he did was meld some clear and refreshing ways of thinking about mapping and GIS to a nicety of empirical sensitivity. There is no Grand Theory.

Truth and Power: Cartography as a Social Practice

So what did Harley actually say? He was originally a British historical geographer. In his early career he worked on the historical geography of medieval England, and wrote a guide to the Ordnance Survey for historians (Harley 1964). Harley loved the old OS map series and wrote historical bibliographic essays for many of the "1-inch" sheets when they were reissued in facsimile (Harley 1969–71). He realized that maps were a critical source material for his work, and developed methods for evaluating them. For if historical maps were not reliable, or only reliable in certain respects, this would obviously impact the historical conclusions that one might want to draw from them.

During the 1970s Harley often met David Woodward, and perhaps neither of them could anticipate how momentous was to be their collaboration. Not only did the two men establish the ongoing History of Cartography project, which is perhaps the most significant intellectual project in cartography, but Harley was galvanized by the sheer diversity of maps to rethink his definition of the map. The shift is clear and involved a move from methodology to a more theoretically informed approach.

In a remarkable series of papers during the 1980s and early 1990s Harley sought to challenge the map as a communication device and reinvest it with power effects (Harley 1988a; 1988b; 1989a; 1990; Harley and Zandvliet 1992). As we saw in Chapter 5, the map communication model proved to be insufficient for many writers. If Harley seized on this weakness with élan, he did not have a consistent idea of what to replace it with. Time and time again he struggled to find other understandings of the map and of cartography.

Certainly in perhaps his best-known paper on deconstructing the map (Harley 1989a) he appealed to the work of Derrida and Foucault to situate maps as players in a system of power/knowledge, that were not autonomous from society. In a revised version of the paper (Harley 1992a) he said we should never consider the map to be "above the politics of knowledge" (1992a: 232), a term he drew from his reading of Foucault. But it is not possible to say how far he wanted to go with this, and his death just two years later at the early age of 59 in December 1991 obviously precluded any further explorations of these ideas.

However, it is fairly clear in his writings what Harley rejected. First, that the map was only true because it was scientific. Harley was not anti-science, but he clearly understood that science did not provide the whole truth, and the truths that it did provide were apt to exclude other knowledges (for example from art, or from pre-scientific cultures). Harley on many occasions spoke out strongly against the all too eager efforts of cartographers and GIS to get taken seriously by being scientific. In this he echoed the work of earlier writers, who while they occupied a different position on the intellectual spectrum were equally skeptical of scientific authenticity. As the former AGS librarian and director J. K. Wright had written as early as 1942 "the trim, precise, and clean-cut appearance that a well drawn map presents lends it an air of scientific authenticity that may or may not be deserved" (Wright 1942: 527).

Harley's position was more radical than this however, and it brings us to his second problem. There was no account in GIS or cartography of how maps act within social contexts, something he would later identify as relations of power/knowledge. Harley and Woodward's history would be thoroughly modern, based on new scholarship and written to the highest standards. Designed to do more than just update previous histories, Woodward and Harley conceived a project that would do nothing less than redefine their subject, maps and mapping. Nor would theirs be merely the story of maps as one of the triumph of knowledge and science over unreason, religious dogma, and superstition. With Brian Harley as co-editor the project would become a critical history, in which claims to objectivity and naturalism would be examined, and perhaps more radically, the Western definition of the map expanded to include indigenous and non-Western mapping traditions (some brought to light in the West for the first time). Although not a predominant theme, the series would also not shy away from the political dimensions of mapping, that is, the power relations exercised in mapping territories, either one's own or that of others', and of the kinds of knowledge that are produced in doing so. Maps were understood not just as efficient documents recording the truth of the landscape, but as active instruments in the very production of that truth (Edney 2005a).

The Peters Map Controversy: Situated Knowledges and the Politics of Truth

". . . the evil Mercator."   J. Paul Goode, 1922, former President of the AAG

One of the most pertinent examples of the Harleian emphasis on maps as situated and political knowledges dwells at the heart of cartography itself. It involves a long controversy over a map made by a German historian, Arno Peters, that split the discipline, and catapulted an obscure Scottish clergyman to notoriety.

In order to see why, we need to fly north from Devon to the historical city of Edinburgh. At 65 High Street in Edinburgh's Royal Mile stands the Carrubber's Christian Centre, an evangelical church. It's one of the few buildings on the Mile to still be in use for its original purpose – worship. The church was founded in 1858 by four leading evangelicals of the day. When one of the founders died in 1895, a man named James Gall, the yearbook of the church recorded these words:

> Our loved brother was from his youth devoted to the Lord's work. . . . A cultivated mind, a fertile imagination, and a thorough knowledge of his Bible made him an able, interesting writer and teacher, though some may think his imagination in his books occasionally got too much scope. (Carrubber's Mission 1983)

If his own church's obituary writer found Gall's writings to be too imaginative, it is easy to see why. Although among his writings there are several books on astronomy, there are others which discuss the origins of "primeval man" as "an anthropology of the Bible." This latter was an attempt to reconcile the new archeological findings emerging in the mid-nineteenth century, Darwin's theory of evolution, and the author's own religious doctrines. According to Gall this reconciliation could only take place if we accept that humans existed *prior* to Adam, a doctrine known as "Preadamism." Preadamites accepted an old earth in accordance with the geological findings of the nineteenth century, but also retained a literal belief in Biblical catastrophes, divinely inspired scriptures, and the historical actuality of Adam and Eve and their descendents including Noah (Livingstone 2008). Gall also described Jesus as living on today in material form "in some constellation," having risen in the Ascension and literally taking physical presence in the heavens above!

How did Gall go from these arcane beliefs to becoming a central and favorite authority figure for cartographers? The answer to this question will illustrate how geographical knowledge is situated in particular times and places. James Gall is often lauded by cartographers as the actual original inventor of Arno Peters' projection, proposed in the late 1960s and early 1970s. Information about Gall and his beliefs is very much lacking even in some authoritative texts (much of what follows is presented for the first time). But if Gall was an extraordinary man with some fantastic beliefs, he was not a cartographer. Yes, he claimed that we could see through the crater of Aristarchus on the moon to the people living in the moon's hollow interior, lit by an incandescent nucleus in perpetual day (Gall 1860[1858]). And he claimed that a race of men lived before Adam and Eve were expelled from Eden 6,000 years ago, a race what's more which was descended from the fallen angel Satan. But even these ideas have their place – and maps were part of it. Cartographers who invoke him as one of their own are missing the real story.

Preadamism

Preadamism is little known today, or for that matter a hundred and fifty years ago, yet it was a sincere and often clever attempt to reconcile emerging scientific findings with the divine word of the Bible (Snobelen 2001). It shares some elements with modern creationists in that the people and events described in the Bible were accepted as real historical characters (Adam, Eve, Noah), that the Bible was inerrant, and that it was divinely inspired. Gall extended the theory of Preadamism in his own way by explaining that the first five chapters of Genesis were not written by a divinely inspired Moses (known as the Pentateuch; the first five books make up the Jewish Torah and were ascribed to Moses by the early Church), but that Chapter 1 was written by Noah about a thousand years *after* Chapter 2, which was written by Adam based on his personal experiences.

While Gall's stated agenda was to reconcile the Bible and science, he saw no reason to reign in an imagination that at times verges on the science fiction of Jules Verne. Putting himself in the position of one of the moon's inhabitants, Gall says:

> What a delightful field for speculation does this present to our friends, the advocates of the universal habitability of the stars! We have been looking to the wrong side of the moon for its inhabitants, and mourning, very unnecessarily, over a scanty, or rather an absent atmosphere. Let us take a peep within, and then what do we see? A concave world – where gravitation is so equally divided, that, while we walk the inverted crust, we may almost choose to fly up into its cloudy atmosphere. . . . We live under the smile of a perpetual day, and enjoy the warmth of a *very* decided summer. (Gall 1860[1858]: 50–1)

In imagining this inner world below the moon's crater Aristarchus, Gall is drawing on a long tradition of hollow earth narratives. Verne's famous *Journey to the Center of the Earth* (1863, translated into English 1872), which appeared shortly after Gall's book, is more accurately described as a deep cave book. But there were the seventeenth-century speculations of the astronomer Edmond Halley, who postulated in 1692 that the earth as well as other planets had a sort of nested set of inner worlds within, at the center of which was a small sun. These were later retold and popularized by a US soldier who had fought in the War of 1812, John Symmes, and were very influential throughout the nineteenth century (Clute and Nicholls 1995).

Gall's speculations are no doubt strange to us today (in the next chapter he went on to discuss the probability of intelligent life on the *sun*), which makes it all the more interesting why cartographers should have latched so deeply onto Gall in the first place, and so assiduously denigrated Peters. There is actually much that the two men have in common. Both were deeply religious and inspired by a desire to improve the world – indeed the Peters map was taken up and distributed by many evangelical organizations. Gall would have found that appropriate.

James Gall, Minister of the Church of Scotland

James Gall was born in Edinburgh on September 27, 1808, to a family of Edinburgh printers and map publishers (Figure 7.3). Both he and his father (also named James Gall) were evangelists somewhat similar to today's Christian Scientists. Another way of describing him is a scholar who sought the hand of God everywhere in nature (Livingstone 1992a: 105ff.).

A major reason why Gall is interesting to us today is the manner in which he sought to integrate scientific and religious worldviews. Rather than rejecting the emerging findings from geology and Darwinian evolution as today's creationists do, Gall saw them as mutually reinforcing:

> If we, by a kind of stereoscopic vision, can see the same objects with the scientific, and at the same time with the historic eye, and find both representations to agree, they would stand out before us with a distinctness, and a visible reality such as they could not otherwise assume . . . the scientific Christian may know and understand things connected with the Bible, which the unscientific Christian cannot know; and the Christian philosopher may know and understand things connected with science, unintelligible to the man who has not studied his Bible. (Gall 1860[1858])

Rev. JAMES GALL,
Founder and President of the Carrubber's
Close Mission.

Figure 7.3 The Rev. James Gall. Source: Annual Reports of Carrubber's Christian Centre. Used with Permission.

In other words, the Bible is a historical document and recounts events that took place on the earth; an earth also illuminated through science. The two work hand in hand. The idea of the Two Books, one of nature and one of God, which could each be read for their truth, can be traced to the Bible itself (Psalm 19) and to Francis Bacon.[1]

In his book Gall explains that his interest in astronomy was derived from the fact that the heavens are the literal dwelling place of saints, angels, devils, and ultimately God, Jesus, and the Holy Spirit. For Gall, when the astronomer reveals something of the nature of stars or other planets, the Christian scientist "can inform him that we have already historically made their acquaintance" (Gall 1860[1858]: 25).

Gall's work on astronomy in turn led him to think about mapping and projections. He explains in a small star atlas he later published that it is very difficult to represent the night sky in anything like a complete form. We need some sort of panoramic projection "including three-fourths of the heavens" (Gall 1885: 119), rather than small sections. His problem was made harder by his wish to retain the form and area of the constellations, so that they looked the same as they do to an observer from earth. He first tried the Mercator projection, but it did not suffice. Gall writes that "[i]t then occurred to me that if, instead of rectifying the latitude to the longitude throughout, [as in the Mercator] we rectified it only at the 45th degree" a compromise projection could be formed that while neither equal area or conformal, distorted area and scale much less at higher latitudes (Gall 1871: 159). In other words, the spacing between the lines of latitude toward the poles does not increase as fast as on the Mercator (which gets its great areal distortions due to adjusting – or "rectifying" – the latitudes in proportion to the amount that the converging meridians have been straightened). Gall's projection was a modification of the cylindrical projection, and it is a secant projection, that is with two lines of true scale (at 45 degrees north and south of the equator; where there is one such line, e.g., the equator, it is known as tangent). Gall called his projection the "Gall Stereographic."

Gall's next step was to apply his new projection to the world, which Gall found "a great improvement on Mercator" (1871: 159). In 1855 Gall attended the Glasgow meetings of the British Association for the Advancement of Science (the BA) where he gave two papers. His remarks were published in 1856 in the BA's official account of the meeting (Gall 1856). In the first paper Gall briefly described his stereographic projection as well as two variants, an isographic and an orthographic. All three projections are cylindrical, and have standard lines at 45 degrees, and it is the orthographic that Peters also later developed. The orthographic is equal-area, with two lines of correct scale, again at 45 degrees north and south. It is this orthographic projection that is identical to the one Peters developed in the 1960s.

James Gall, Preadamite

Gall's life intersects the after-effects of the Scottish Enlightenment and the challenges it posed to forms of knowledge. Where previously generations of writers had been satisfied with ecclesiastical scholarship this was no longer the final arbiter of knowledge. If fossilized amphibians were discovered high up in the mountains (as James Hutton had done on the Salisbury Crags in Edinburgh in 1770), then it was possible that the earth had gone through a long upheaval. It was part of the increasing evidence that the earth was not young, raising the probability that strict Biblical accounts of the creation would have to be discarded. The members of the Scottish Enlightenment desired a new social order based on human rationality and morality. But at the same time they had no wish to *discard* religion. In the new hoped-for scheme of things, humans and human reason were the fundamental sources of morals, thought, and civil society. Science and God would both exist, but the question was their relationship. Did one supersede the other? Were they mutually reinforcing? Could the Bible be made to square with scientific evidence?

Gall's answer came in the theory of Preadamism. Gall conceded – and even emphasized – that the Bible makes no scientific predictions, and therefore it is "a book for all men in all ages" (Gall 1860[1858]: 21; 1880: 13). He continued, "if the Bible, therefore, made any revelations whatever in such a field, it would have betrayed a human origin" (1860[1858]: 21). But the reader of the Bible is best positioned to carry out scientific observations. The scenes of the Bible:

> are not laid in fictitious localities, nor are its events unconnected with the world as we find it. On the contrary, it is linking itself with every known fact more strongly and more distinctly every day; and history, geography, and even other sciences, are continually bringing up fresh evidences of its entire truthfulness. (Gall 1880)

In particular, astronomy can contribute to our knowledge because Christ is incarnate; He has a material body. Christ's Ascension has a physical residence in the heavens and the "astronomer may have already pointed his telescope unwittingly to the very star where at present dwells the Judge whose voice he will yet hear" (1860[1858]: 25).

Gall argued that the creation story in Genesis did not literally refer to six days of creation. In order to unify the testimony of scripture and geology, Gall advocated the "gap theory" of Preadamism (Livingstone 1992b). Preadamite gap theorists cleverly reconciled Genesis with fossil evidence and science by postulating a long time gap between Genesis Chapter 1 (the creation of the earth in six days) with Genesis Chapter 2 (the story of Adam and Eve). Gall explains:

> the Bible narrative does not commence with creation, as is commonly supposed, but with the formation of Adam and Eve, millions of years after our planet had been created. Its previous history, so far as Scripture is concerned is yet unwritten. There may have been not one, but twenty different races upon the earth before the time of Adam, just as there may be twenty different races of men on other worlds. (Gall 1880: 50)

Gall's writings on these matters underwent some interesting changes. These changes hinge around Darwin's *Origin of Species* (1859). Before Darwin, Gall was clearly unsure how to reconcile the Bible with scientific findings, but already recognized that the six-day account of creation could not be literally true. He was unsure in the sense that he believed science and religion were in accord, but that the exact details were unclear. After Darwin and the new geological evidence of the mid-century, Gall expanded and refined his earlier work (Gall 1880: first edn. 1870). In other words Gall, as a biblical believer, was actually *assisted* by the Darwinian theory of natural selection. He took advantage not only of Darwin, but Lyell (1863) and Preadamite writings.

Gall asserted that the earlier remains in the geological record are of Preadamite races that had exhausted themselves before Adam. Gall leavened this typical Preadamite belief with some rather unusual additions, even for Preadamite works of the time. He maintained that Preadamites were descendents of angels, and that the pre-fall Lucifer was their father. Gall was very interested in Lucifer or Satan and saw him as an ordinary if long-lived man who had never been in heaven. These Preadamite races sinned and fell into ruin, and evidence of their primitiveness is found in the "Danish shell-heaps, the Brixton [sic] caverns, and the mud of the Nile" (Gall 1880: 27), that is, what we would see as prehistoric archaeological evidence.[2] All this seems far from cartography – but is it?

The Peters Map

Certainly, some 115 years later when Arno Peters unveiled his map at press conferences he surely was not thinking of James Gall. Rather, according to one journalist Peters said that he had created the projection so that every country would be in its exact relative size to one another. He had set himself the task, he said, of using a rectangular projection because "we live in a four-cornered world, [and] the television tube we sit in front of is perhaps the best symbol of it" (Morris 1973: 15). Although the Mercator projection is also rectangular and had its uses, Peters was very critical of it. He described it as presenting "a fully false picture, particularly regarding the non-white-peopled lands . . . it over-values the white man and distorts the picture of the world to the advantage of the colonial masters of the time." For Peters the Mercator gave an unreasonable amount of space and therefore importance to the smaller Western countries. On the other hand large countries outside the West are shown relatively small. This disparity did not equate with the true size of these places. Monmonier has joked that Peters was concerned not so much with fairness to all peoples as "fairness to all acres" (Monmonier 1995: 39). (See Figure 7.4.)

Cartographers were at first only mildly interested in the projection. But then two things happened that raised their ire. First, the map started to become popular – very popular. To date an estimated 80 to 85 million copies of the map have been circulated (Devlin 1983; *Economist* 1989). Second, Peters attached a number of claims to the projection that cartographers took objection to, most significantly in his book *The New Cartography* (Peters 1983).

Here's a fairly typical example of the cartographic reaction. Arthur Robinson, Peters' most energetic critic, eloquently described Peters' book as a:

> cleverly contrived, cunningly deceptive attack against the "outmoded theories" and "myths" of cartography [it] misrepresents, is illogical and erroneous, and one's initial reaction is simply to dismiss it as being worthless . . . [Peters is a] skilful merchandiser, and his self-serving campaign can do the image of cartography great harm. (Robinson 1985: 103)

Aside from the critical response to Peters' cartographic claims, the single most enduring rebuttal is that the Peters projection is not even original. Cartographers have often turned to Gall's work on projections as the original and prior work. As we have seen, Gall did indeed publish a projection that to all intents and purposes is the same as the one that Peters developed. Yet asked whether he was familiar with the Gall projection in 1989, Peters said he was "not aware of the Gall variation until recently" (Monmonier 1995: 27).

Attack articles such as this one made easy meat of Peters, and within cartography at least, opinion was settled. The controversy died down in the 1990s, but it flares up from time to time. Recently it received new publicity in an episode of the TV series *The West Wing*. As one author has noted, the "Peters phenomena" will simply not go away (Vujakovic 2002).

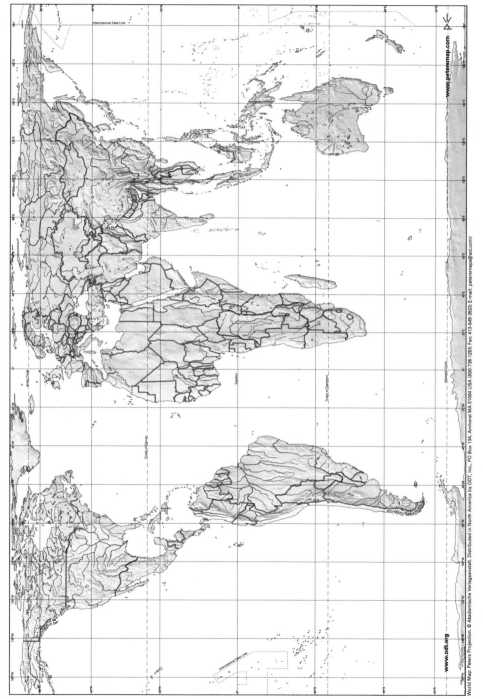

World Map: Peters Projection. © Akademische Verlagsanstalt. Distributed in North America by ODT, Inc., PO Box 134, Amherst MA 01004 USA (800-736-1293; Fax: 413-549-3503; E-mail: petersmaps@aol.com)

Figure 7.4 The Peters Projection or World Map. Source: www.ODTmaps. © 2005, Akademische Verlagsanstalt.

Peters further explained in an interview toward the end of his life:

> I want a map where the sizes are exactly the same as in reality. . . . And so we don't
> need a new map; we need a new view of the world . . . I [found] out that the world-
> view is what I'm looking for, not the map. The map only is the possibility to give a
> worldview. (ODT Inc. 2008)

The "new view of the world" became Peters' catchphrase. In other words, like Gall, Peters saw the map as a means to an end. Peters suggests that the way we see and understand the world, including its geography, influences our actions. Foreign policy and aid decisions for example, flow out of our global perspective. Military build-ups occur where threats are perceived (we might compare here the idea of the "axis of evil").

Throughout his life, Peters was involved in socially progressive causes. One of his earliest political memories was the visit of William Pickens in 1929 when Peters was 13. Pickens, whose parents were liberated slaves, was a graduate of Yale and Vice-President of Morgan College in Baltimore. He had been involved in the NAACP since its inception in 1910 and was probably America's best-known black man and later president of the NAACP. A fluent speaker of German, Pickens traveled to Germany at least four times (Avery 1989). He expressed much admiration for the resistance of ordinary people to the increasingly authoritarian rule of the Nazi party. Pickens inscribed a copy of his autobiographical book *Bursting Bonds* (Pickens 1991/1923) to Lucy Peters, Arno's mother (ODT Inc. 2008).

As a historian Peters argued that Europe-centered bias was no accident but was rather a deliberate geo-political policy:

> This geographical view of the world is designed to eternalize the personal overestima-
> tion of the white man and in particular the European while keeping colored peoples
> conscious of their impotence . . .
>   [The Mercator] map is an expression of the epoch of the Europeanisation of the
> world, the age in which the white man ruled the world, the epoch of the colonial exploita-
> tion of the world by a minority of well-armed, technically superior, ruthless white
> master races . . . . (Peters 1974: n.p.)

In other words, Peters argued that maps, and especially the well-known and at that time still often used Mercator map, were products of their time, and that they were complicit in reproducing geo-political exploitation. What is remarkable today about this argument is just how far ahead of his time Peters was. During the 1950s and 1960s when Peters began work on his history and his map, there is almost no mention in the geographical literature of the role of geographical knowledge (such as maps) in power relations and exploitation. As we saw in Chapter 2, whereas anthropology and the Marxist tradition had long documented these political conditions, geography and political geography were struggling to shake off the early twentieth-century doctrines of environmental determinism and Haushofer's geopolitics (Agnew 2002). In cartography's case, these arguments would not be taken up again until nearly 20 years later with the work of Brian Harley.

Harley in Person

I was fortunate enough to meet Harley two times, both in 1991. As a graduate student at Penn State, we often benefited from a unique lecture series enabled by the late Peter Gould, who was a named professor. Gould generously used the honorarium that came with this position to fund a number of speakers in a Distinguished Lecture series. In early 1991 we graduate students approached Gould with the suggestion that he ask Brian Harley, then as we thought an emerging young and radical scholar (for so he seemed from his writing), to come and visit. Luckily Gould agreed and in March Harley was able to visit and deliver four lectures that he planned to make the basis of a forthcoming book he was working on with John Pickles called *The Map as Ideology* (the title of his second lecture, see Figure 7.5).

It was the tradition at these visits for the speaker to spend one evening, and one evening only, hosted by the graduate students with strictly no faculty present. Since Harley was basically the suggestion of myself and John Krygier (now at Ohio Wesleyan University) it was decided to hold the party at my place. By now we'd got over the shock that this Foucault- and Derrida-quoting scholar was not, as we thought, our own age, but 59 years old!

Eliot may claim that April is the cruelest month, but in central Pennsylvania March is a close contender. It was freezing. Snow was still on the ground and my small apartment on Circleville Road was too small to contain everybody. After a few drinks to warm us up, we piled outside and started throwing snowballs – including our Distinguished Speaker!

The last and second time I saw him was coincidentally at the end of the AAG meetings that year the next month. We were both in line to check in at the airport – he to Wisconsin, myself to Pennsylvania. We wished each other a safe flight as you do and got on our flights. In September of that year I had to return to the UK (basically my funding had run out and Penn State had rightly got tired of funding me). It was Christmas and I was sitting on my father's settee upstairs in his living room reading the newspaper. Suddenly a familiar (though younger) face appeared, that of Brian Harley. For a moment I didn't understand why I should see his face in the newspaper.

The following March I was able to attend his memorial service in London, held at the Royal Geographical Society in Kensington. It was a short train ride from Portsmouth, where I was then living, though I had been warned that that route was pretty lousy. At the RGS I met Harley's three daughters and listened to many tributes to his work from former colleagues in the UK and America (Royal Geographical Society 1992).

I add these biographical details, not out of any claim to "know" Brian Harley or to put something "sweet" in (as a reviewer of the manuscript claimed), but because of a key piece of writing that is virtually unknown today (it wasn't included in his posthumous "best of" collection). It is called, simply "The Map as Biography" and it was not published in an academic journal but rather a map collecting magazine.

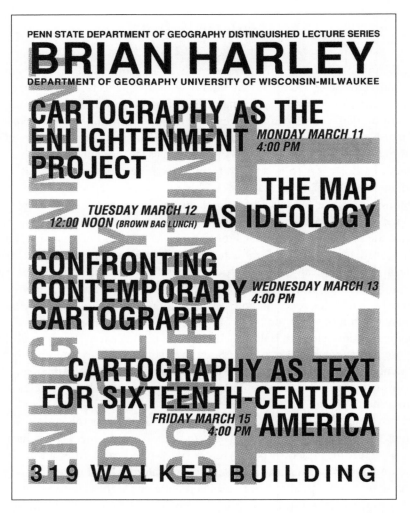

Figure 7.5   Flyer for Harley's 1991 talks at Penn State (designed by John Krygier).
Source: John Krygier. Used with Permission.

This piece is key if you want to understand Harley – and it's only three pages long! I ask my students to read it every year, and reread it myself, and I'm always taken again by its power.

   In the piece Harley talks about the way maps can evoke meaning: "the map has become a graphic autobiography; it restores time to memory and it recreates for the inner eye the fabric and seasons of a former life" (Harley 1987: 20). *It restores time to memory.* Harley had recently moved from England to Milwaukee and the map he chose was one that intersected with his own life in many ways: it was a map of six square miles centered on his old town of Newton Abbot where the "Devon walk" took place. Harley notices all sorts of details over the map, "personal experiences and cumulative associations give to its austere lines and measured alphabets yet another set of unique meanings. Even its white spaces are crowded with thoughts as I whimsically reflect on its silences . . . here is the lane too where – not long ago – I met a woman on a summer evening: the overgrown wall of her orchard is marked on the map" (20).

De Certeau of course talked about a philosophy of the everyday life, of those myriad small moments, movements, trajectories, and wanderings that actually comprise our lives (de Certeau 1984). But Harley was no philosopher, even less was he a "postmodernist" as is sometimes absurdly claimed (Ormeling 1992), but really the most modernist of men. If during the seventeenth and eighteenth centuries the map was a mirror, and if for postmodernists (if they exist) truth is an illusion, then for the modernist of the twentieth century such as Harley truth is buried beneath illusion.

Harley believed that the truth in maps was essentially present, buried as it was beneath lies, propaganda, and interests that had accreted around it. That was Harley's "deconstruction" then; not that of Derrida, but the recovery and questioning of true meaning. And you might well ask, how far is that from what Arthur Robinson sought? Not quite so far, after all.

This little article on the map as biography shows that such a search does not take place in the realm of the theoretical, but in the practical material world of map sheets and atlases that are carried with you (Harley had a well-known incompetency when it came to computers and "evil-mail" as he called it).

The final thing the article does, in what might be its most powerful act, is to show the beauty in this view, and that the act of *finding* meaning in its situated places (and the mapping devices used to do that) is itself a form of beauty.

Notes

1   I thank David Livingstone for this reference. It was popular among nineteenth-century evangelicals such as Gall and Isabelle Duncan (Duncan 1860/1869; Snobelen 2001).
2   The Brixham cave near Torquay in Devonshire discovered in 1858 gave strong new evidence of the co-existence of human occupation alongside now extinct animals, and therefore was evidence of human antiquity. The Danish shell mounds were refuse mounds of consumed shellfish, mixed with bones of other animals and stone-age implements. The alluvial plain of the Nile was excavated between 1851–4 by the Royal Society. All these examples were drawn by Gall from Charles Lyell's *The Antiquity of Man* (Lyell 1863).

Chapter 8

GIS After Critique: What Next?

"GIS does not simply 'visualize' data, it has an ontological power."

Marianna Pavlovskaya (2009)

The GIS Wars

On the front page of the AAG newsletter in 1988 the then-president of the association, the cultural geographer Terry Jordan, wrote a short piece called "The Intellectual Core" about the discipline of geography. It was Jordan's last column before stepping down as president and he made sure that it was a memorable one. Jordan worried that GIS was swamping out traditional specialties (such as regional, cultural, and historical geographies – his own areas). GIS was, in Jordan's words, an easily justified but "nonintellectual" activity that might guarantee jobs but did not amount to scholarship (Jordan 1988).

This was perhaps the first shot in the so-called "GIS wars." It would not be the last nor even the worst. The newsletter is sent to every member of the AAG and it caused widespread resentment among GIS users. Everyone is entitled to their opinion, but this came from a sitting AAG executive! Meanwhile other people began to voice their disquiet about GIS. For Taylor GIS was a turn away from valuable knowledge to fact-based "information."

For others GIS threatened a return to positivism and the surveillant society (Pickles 1991). In perhaps the harshest comment it was asserted that GIS had led to the "killing fields of the Iraqi desert" (a reference to the Gulf War of 1990–1) (Smith 1992). And was it also killing off cartography?

This latter was a widely held fear. If the love of maps is still the love that is hard to confess, few doubted that cartography wasn't a core component of geography. By the time cartographer Judy Olson took over the presidency of the AAG at her plenary session at the 1996 meetings in Charlotte, North Carolina, she could pointedly ask the question "Has GIS killed cartography?"

On the other hand there seemed to be an equal and opposite reaction against "postmodern" social theory as simultaneously dangerous and irrelevant. Critical social theory is "essentially destructive and individualistic" (Openshaw 1997: 8). The well-known cartographer Mark Monmonier reportedly remarked that postmodernism lead to MAD – mutually assured deconstruction (cited in Openshaw 1997: 24).

Openshaw was perhaps only the most vocal at claiming that positivism was not only desired but that it represented our (geography's) best hope to become a science. In a now-famous quote, he wrote in somewhat puckish mood:

> A geographer of the impending new order may well be able to analyse river networks on Mars on Monday, study cancer in Bristol on Tuesday, map the underclass of London on Wednesday, analyse groundwater flow in the Amazon basin on Friday. What of it? Indeed, this is only the beginning. (Openshaw 1991)

To which some may respond: "I'd still be labeling Tuesday's cancer polygons on Friday" (Stacey Warren, cited in Schuurman 2000).

Others were angrier. Copies of Pickles' book *Ground Truth* were waved around at an IBG meeting with the comment "this is a manifesto of destruction" (cited in Schuurman 2002: 294). GIS was eloquently defended. It could reunite the discipline – put "humpty-dumpty together again" (Openshaw 1991). GIS was misunderstood and did social critics even use it? Michael Goodchild (who coined the term "Geographic Information Science" in 1992) and a leading GIS expert once remarked apropos GIS as the return of positivism: "It's amazing for me to read that positivism died in the 1970s. From my perspective, it's not even been ill" (Schuurman 1999b: 4).

From today's perspective these comments seem well past their sell-by date. Today we have a $10bn a year GIS business, the geoweb and Google Earth, feminist GIS (Kwan 2002a), GIS being used for human rights (Zetter 2007), and burgeoning demand for GIS in the classroom. What happened? Did GIS "win"? Where does this leave critical GIS today?

A Short History of the GIS Wars

The particularities of this debate have been traced in useful articles by Nadine Schuurman (Schuurman 2000; 2002). She characterized the debate as having two major sets of players: those from a broadly social theoretic position, and those using or researching GIS. Schuurman describes the encounter between these two groups as having three different rounds of engagement starting in the late 1980s.

In 1991 at the AAG annual meetings in Miami there were numerous sessions on "Ethics in Cartography" and "Space, Power and Representation: Social Action and the Socio-Spatial Dialectic" organized by Mark Monmonier, John Pickles, and

Table 8.1 Periodization of the GIS and Society debate.

1988–1992	A round of criticism, attack and accusation in which GIS was labeled "nonintellectual" and positivist, as well as some robust counter-attacks from GISers. Example literature: Openshaw 1991; 1992; Taylor 1990
1993–1998	A second round of increasing sophistication from critics, meetings between the two groups and the launching of a special initiative within the NCGIA. Example literature: Pickles 1995; Sheppard 1995
1998–2001	In the third round, both GIS critics and GIS researchers recognize the non-deterministic flexibility of GIS. Both "sides" are increasingly leaving the debate behind to produce more socially responsible or "reconstructed" GIS. Example literature: Elwood 2006a; Kwan 2002a; St. Martin and Wing 2007

Source: adapted from Schuurman (2000; 2002)

John Paul Jones. Among those giving papers were Patrick McHaffie, Sona Karentz Andrews (a Milwaukee colleague of Brian Harley's), Michael Curry, Wolfgang Natter, and Ed Soja. These sessions took their point of departure from Brian Harley's "ethics" papers (Harley 1990a; 1991) which had attempted to challenge and rethink the notion of ethical mapping by considering issues of power, politics, and knowledge. (Harley also acted as discussant.)

Despite some fairly substantial criticisms of GIS that were voiced at these sessions (and in an article by Pickles that accused GIS of returning to positivism [Pickles 1991]), it became clear that the majority of participants were willing to engage in substantive dialog. In the spring of 1993 Tom Poiker (then at Simon Fraser University) sent out an invitational email via the GIS listserv. On behalf of a steering committee (Poiker, Eric Sheppard, Nick Chrisman, Helen Couclelis, Michael Goodchild, David Mark, Harlan Onsrund, and John Pickles) participants were invited to a workshop to discuss differences and mend bridges.[1] Known as the "Friday Harbor" meetings after the location on San Juan Island, Washington, were held November 11 to 14, 1993, and hosted the two "sides" in the debate. Scholars with social theoretic interests joined with users and developers of GIS. Friday Harbor was sponsored by the National Center for Geographic Information and Analysis (NCGIA) who also provided funds. Twenty-four positions papers were presented.

The proceedings and discussions at Friday Harbor have never been published as such, but nevertheless this stands as an important moment in the critical engagement between social theorists and GIS. A number of the papers from the retreat did eventually appear in a special issue of the journal *Cartography and Geographic Information Systems* (CaGIS) as well as Pickles' edited collection *Ground Truth* (Pickles 1995).

Tom Poiker later described how the meeting arose:

The preparation of this meeting started with an article by John Pickles (1991). After reading it during Summer 1992, Tom Poiker wrote a letter to John [Pickles], suggesting that the topic needed discussion between social theorists and GIS people. Before John could answer, Tom presented the issue to a group at the University of Buffalo in September 1992 and Dave Mark suggested to submit a proposal to NCGIA for the organization of a workshop on the topic. A letter by Eric Sheppard to Mike Goodchild [who was too ill at the time to attend Friday Harbor] on GIS issues for the social sciences helped the development. (Poiker, report, p. 1)

At the end of the 1990s, Pickles summarized the main points that emerged at Friday Harbor and beyond. The summary represents what "critical GIS" means today:

- to contribute to a theory of GIS which is neither technical nor instrumental, but locates GIS as an object, set of institutions, discourses, and practices that have disciplinary and societal effects;
- to show how these disciplinary and societal effects operate;
- to push against the limits of GIS and its unacknowledged conditions and unintended consequences of development and practice (e.g., corporate influence, epistemological assumptions, and understanding of appropriate applications);
- to ask whether GIS could have been different, or in what ways it may be made different in the future (Pickles 1999: 50).

What Have We Learned: After Critique

After the Friday Harbor meetings faded away a new term started to be used to refer to the work it had started: critical GIS (Schuurman 1999a). This term has some problems (see Chapters 2 and 4 for more discussion of critique); as a number of people have pointed out, it implies that other work is *un*critical. Blomley (2006) tells the story of a senior colleague who objected to its use because surely everything we do is critical? If by critical is meant something like taking care or questioning, then this point seems hard to deny. We are forced to ask then: what have we learned?

Positionality

One of the first things that became apparent was that GIS and the knowledge that could be produced with it had positionality. That is, it had a two-way relationship between itself and society. While this idea became fairly well accepted in newer human geography following the cultural turn of the 1980s, there has been more of struggle to situate geospatial knowledge in a particular time and place – it is still assumed to be "timeless." For example, during the great Peters projection controversy

(Chapter 7) many of the objections from cartographers centered around Peters' more political claims, which they saw as inappropriate to the subject of mapping: there should be no interface at all between mapping and society. Mapping was post-political.

In rebuttal, Sheppard offers the example of the development of the Mercator projection (Sheppard 1995). While the projection "met the societal need in the sixteenth century to improve the ability of European sailors to chart courses and coastlines" it was simultaneously "part of a cluster of cartographic, navigational, and ship-building innovations . . . stimulated by the desire of western Europe's nascent nation states to expand their wealth and influence through mercantile expansion" (1995: 6). In other words, mapping is part of a socio-political nexus of power relations.

Another example of the relation of mapping to wider societal developments may be the way that corporate "Big GIS" such as ESRI ArcGIS is quickly being modified to be compatible with developments in the geoweb (see Chapter 3). This is a case where GIS, rather than being at the forefront of innovation, is struggling to maintain its relevance. According to Google its Google Earth software has been downloaded 400 million times; recent figures from ESRI claim its ArcGIS products have an installed user base of one million. Perhaps GIS is no longer the innovation driver here?

But that's not all. Critical GIS also questions the positionality of GIS and its effects on society. Here we run into the question of what kinds of knowledges are produced with what effects? In Chapter 11 I examine this question more deeply by looking at the way that mapping has constituted knowledges of race and the political effects that this has had. Other examples abound of GIS being used to intervene to redress social inequalities such as environmental justice and neighborhood empowerment using participatory GIS (PGIS) (see below).

This dualistic relationship between technology and society has become a standard way to situate technology. For example, a leading text on technology and modernity sees the relationship as one of mutual "co-construction" (Misa et al. 2003). Technologies such as GIS are structured, and in turn restructure, the nature of modern societies.

Reflexivity

For Sheppard it must mean critique of the self or reflexivity (Sheppard 2005; 2009). Sheppard draws on the Frankfurt School to connect GIS practices to a wider field, especially critical human geography. Do not expect consensus, he warns: "we continue to be insecure about what 'critical' means" (Sheppard 2009), and productively so. One can also be reflexive about what happens to local knowledges if and when they are incorporated into a GIS. Rundstrom has argued in this context for a politics of refusal "of not telling, not writing, not encoding" (Rundstrom 1995: 53).

PGIS democratic possibilities

For Pavlovskaya (2006) critical GIS means overcoming current practices:

> "Critical" clearly implies questioning the status quo, whether it is dominant know-ledge production practices or dominant configurations of social power. It is also going beyond critique and thinking about how possibilities, creating new social imagina-tions, and producing hope in and desire for those imaginations. Critical GIS, then, is a field that conceives of how geo-spatial technologies can be used to counter scientific and social conservatism. (Pavlovskaya 2009)

For Pavlovskaya it is not sufficient to just perform critique; although that step is necessary to produce the social histories of GIS there must still be something "after critique." For her, this means explicitly "progressive social research that is . . . informed by critical human geography perspectives" that aims to bring about a "postpositivist sensibility" (Pavlovskaya 2009). Whether positivism can really be equated with quantitative methodologies is a moot point and is often denied by quantitative geographers (Schuurman 1999b). However, Pavlovskaya points out that the quantitative possibilities of GIS are actually quite limited, and that GIS is open to a wide variety of qualitative analytical techniques such as ethnographies (Pavlovskaya 2006). Certainly for many people the notion that GIS is not primarily quantitative will challenge their preconceptions. But what Pavlovskaya is arguing is a good example of what can come after critique: the resituating of a technology. GIS is a powerful approach, she says, for visualizing new possibilities. Like maps themselves, GIS is a powerful means for visual persuasion and its geographical imagination can reveal or create alternative worlds: "GIS does not simply 'visualize' data, *it has an ontological power*" (Pavlovskaya 2009: emphasis added).

Similarly, Dunn (2007) explores how local and indigenous knowledges may be incorporated into a participatory GIS, or what might be called "a people's GIS." Dunn identifies four issues for PGIS:

1. Who controls and has access to information?
2. How are local and indigenous knowledges to be represented?
3. How can GIS be democratized?
4. Sustainability and long-term impact.

She notes that there is no single answer to these questions. For example, with reference to how local and indigenous knowledges may be represented, questions here may include whether it is possible at all (which some commentators have long denied, see Rundstrom 1995), and how (or whether) to integrate such knowledges with "expert" or official information. Folklore, proverbs, songs, and stories: how would these be "represented" and if they are, what value would they have? This last point has been raised several times by other writers who note that it is all too possible for local knowledges in PGIS to find themselves being restructured so that they "fit" with elite knowledges. For example, if there "are different notions of precision and 'accuracy'" how might local knowledges be integrated with other spatial data (Dunn 2007: 622)? Would they be cleaned up, deleted, or transformed? Since to date the vast majority of GIS are based on a Cartesian view of space, a non-Cartesian one may be incompatible (this point is often explored in the work of map artists who overlay incommensurate geographies, see Chapter 12).

Furthermore, just because local communities are involved does not mean that they are in control; indeed they may simply be providing a cover for exploitation to continue (cooptation).

All of this sounds somewhat gloomy. In reality, the role of communities is likely to be one of ongoing negotiation, with "total" exploitation or empowerment unlikely to occur. As with the "netroots" discussed in Chapter 3, communities will more likely experience a process of continual engagement. Furthermore, these questions usefully point to the partiality or situatedness of all knowledge, *including* elite or expert knowledge. Elwood's ethnographic research in Chicago communities for example has uncovered a nexus of relations that are neither cooptation nor pure resistance:

> By producing flexible spatial narratives that enable them to pursue multiple objectives, community organizations strategically navigate the institutional, spatial, and knowledge politics that produce and transform urban spaces, in a way that cannot be solely characterized as either cooptation by or resistance to more powerful state and business interests. (Elwood 2006a: 324)

Feminist GIS

A particularly productive element of critical GIS has been feminist GIS (Kwan 2002a; 2007; Schuurman 2002). If the GIS Wars were reductive, forcing participants to take one of two highly caricatured positions, then feminist GIS has emphasized a more inclusive and contextualized strategy. In rejecting the binary of the typical story's Us versus Them, feminists draw on a legacy of work critiquing gender binaries and gender-related identities. Kwan emphasizes the subjectivity of those who use GIS that may act to undermine simple binary narratives (Kwan 2002b) by recovering the voices of feminist GIS users and researchers. Feminist GIS also tends to place significant emphasis on reflexivity and positionality:

> feminist reflexivities attempt to problematize the relationships among the research, the researcher and the researched, to acknowledge the partiality and positionality of the knowing subject, and to ameliorate the effect of the unequal power relations in academic research. (Kwan 2002b: 275)

In a field that privileges "disembodied" research using secondary data such as remote sensing or the census, feminist GIS seeks less detachment and more use of qualitative data. Kwan draws on Rose's tenets (Rose 2001) for critical visual methodologies to identify three sites where this more reflexive, positioned research can contribute: the site of knowledge production; the site of the representation; and the site of interpretation. All sorts of questions circulate here. On the latter point Kwan asks whether our representations can encourage alternative ways of looking, how the objectifying male gaze may be subverted, and how the production of alternative subjectivities other than that of the master subject may be produced?

In a recent paper Elwood brings us full circle by examining how feminist GIS may be related to the geoweb and to volunteered geographic information (Elwood 2008). Research on the geoweb can benefit, she argues, by taking into account the lessons from feminism and feminist GIS, particularly the idea that GIS is both a social and a technological phenomenon; that an all-encompassing perspective is less appropriate than a flexible multi-vocal one; and that rather than typical models of decision-making that emphasize consensus and well-described data sets, VGI and the geoweb are more likely to face data uncertainty, contradiction, and diversity. Thus if researchers adopt the old model they are likely to run into problems.

These are powerful new considerations for critical GIS that go well beyond what was on the agenda of GIS and Society in the 1990s.

After Critique? Extending Possibilities for GIScience

So what to make of all this? In a widely discussed piece the sociologist Bruno Latour asked what happens when critique becomes so successful that nobody believes anything anymore (Latour 2004). Thus while it's true that most scientists accept anthropogenic causes of global climate change, he cited a report in the *New York Times* about a Republican strategist who urged a lack of certainty in the construction of scientific facts since *some* people were critical of them. If critique is "the victory of uncertainty" then what comes after critique is victorious?

Latour's article is useful for identifying what is at stake here. In a way it is the ultimate act of reflexivity because it challenges the very process of critique itself. What might we say in response? Despite Latour's reservations it is still possible to acknowledge the value of critique and critical GIS.

To begin it does seem as if Latour has fallen into a trap he has warned others about, namely of expecting unassailable grounds for knowledge – that no knowledge is ever finally secure. This lesson is easy enough to apply to knowledges that you personally oppose, but it is a different feeling when they are applied to ones you personally endorse (say, global climate change). Latour seems to be feeling that sense of unease. First, expecting to sort out arguments about global climate change or the truth of the map will not be a matter of appealing to the fact–value distinction or of objective information. Rather, they will be argued at the intersection of science, politics, and power. Knowledge is not beyond politics. This key insight from critique and critical GIS is therefore still valid.

Second, because a truth is understood as being produced, socially constructed, or possessing a historicized genealogy does not make it any less a truth. Rather, it is to have a better account of truth. A produced truth is as real as anything else that is real, such as this desk I'm sitting at. The question then is how was that truth produced? With what effects? And as reflexivity and positionality would suggest, what is your relationship to that truth (do you accept it, oppose it, decry it, etc.)? As an anthropologist might tell you when studying different cultures, it's not that they're all equivalent (as a supposed relativist might say), it's that your embeddedness in a culture structures your world. Different cultures are different worlds.

So for GIS after critique it has to become more diversified to include a critical component as we saw above. But it has also become more nuanced, and prepared to engage on theoretical and even philosophical grounds. Perhaps some of the most interesting work here is in so-called "ontologies."

Over the past several years the term "ontologies" has become more and more noticeable. Arising in computer science, ontologies now pervade the field of GIS and have been deployed in the development of the US National Map. As the signature federal product with the goal of mapping the country's topography (its elevations, boundaries, hydrography, and so on) the National Map is truly a "people's map."

So what are ontologies? One way to think about them is as definitional "triples" in the form of subject–verb–object. For example, "flood" "is a" "weather phenomenon." This might not seem terribly new. In some ways ontologies are what we used to call feature lists or metadata. But their usage in GIScience is apparent if you reflect on how computer systems operate – through stable, well-defined representations of knowledge. GIS won't operate if say "ecologically threatened wildlife" or "NDVI" (Normalized Difference Vegetation Index, a measure of plant health) changes from day to day or user to user. So "ontologies" stabilize knowledge through abstract triples (semantic definitions) to make GIS interoperable. According to one review "geographic ontologies are receiving increasing interest and are growing in significance" (Agarwal 2005: 502).

GIS ontologies are extensible and can contain incompatible information. For example, Rob Raskin of NASA's Jet Propulsion Laboratory (JPL) has developed SWEET, the Semantic Web for Earth and Environmental Terminology. SWEET is an expandable, upper-level set of concepts in the form of ontologies (Raskin 2005).

But GIS is not the only endeavor that employs ontologies. In a recent article in *Scientific American*, Nigel Shadbolt and Tim Berners-Lee promote formal ontologies as part of the new discipline of Web Science (Berners-Lee is the inventor of the World Wide Web). If you connect up all your triples into a network (called a semantic web) information becomes more usable. Each triple can also be given a Uniform Resource Identifier (URI) which means that it can be located on the Web just like a webpage. Other people can use these semantic webs – networks of triples – to find, understand, and access information. Shadbolt and Berners-Lee offer the example of finding a used Toyota car below a certain price for sale in your home-town. A regular Web search (e.g., "second-hand Toyota") will yield too many responses that do not fall within our parameters. With a used-car semantic web, which could deploy a chain of triples such as [car X] [is brand] [Toyota]; [car X] [price] [<$8,000]; [car X] [located in] [your hometown] it can be readily determined that indeed car X is a suitable vehicle (Shadbolt and Berners-Lee 2008).

Semantic webs can also be used to mine other information that has been set up where objects are identified or tagged. If we take the OpenStreetMap (OSM) project for example, all its geographical features have been tagged or identified. OSM is an open-source mapping project that operates under the Creative Commons license. It contains general classes of geographical features, such as "ways" which include things like roads, canals, paths, railroads, and so on. Then each item mapped is tagged more specifically. A road for example might be tagged or classed as a "motorway," "toll road," and "susceptible to traffic jams." This is a semantic web waiting to happen. You (or an automatic software program) could search for specific roads such as all the roads in Atlanta not susceptible to traffic snarl-ups. Since the definitions and tags of OSM data are done by humans (you can define objects as many times as you like), this becomes a powerful search capability. Raskin and colleagues call this semantic web a powerful component of spatial web portals (Li et al. 2008).

There have also been some interesting developments in GIS and narrative, personal space, and cross-cultural comparison of ways of thinking and relating to the environment. For example, David Mark and colleagues have examined categories of thought in indigenous populations in both the United States and Australia (Mark and Turk 2003; Mark et al. 2007). Their work has contributed to our understanding of the different ways that cultures think about geographical terms, and potentially could be used to understand different ways of being.

As a text-based system, semantic webs may nevertheless struggle to encompass non-textual ways of understanding or being. Matthew Sparke documents a trial at the Supreme Court of Canada in which two First Nations (the Wet'suwet'en and Gitxsan) contended against the state over territorial claims. According to Sparke the state's strategy was to "abstract" and decontextualize space in which "cartography enables abstraction away from bodies, social relations, and history" (Sparke 2005: 10). In order to be taken seriously by the state it was therefore necessary to be reconfigured as a "normalized abstract space" (2005: 15). But what of a geographical understanding that is neither textual nor graphical but ceremonial songs *performed* by elders under conditions of semi-secrecy that outsiders are not allowed to see (Turnbull 1993)? How can that world appear in a semantic web, no matter how extensible?

This question has been partially addressed in the work of Kwan and Ding in what they call "geo-narratives" (Kwan and Ding 2008). Geo-narratives are an extension of qualitative and participatory GIS methods. They include oral histories, biographies, and life histories, which have not been extensively incorporated into either GIS or computer-aided qualitative data analysis software (CAQDAS). Their work examines hate crimes against Muslim women following the attacks of 9/11 using a wide variety of qualitative methods which "provides a geographic context that facilitates interpretation and understanding of the lived experiences of the research participants. It allows for the creation and interactive geovisualization of their subjective environment" (Kwan and Ding 2008: 458).

Meanwhile, heading in another (opposite?) direction in the field is research derived from philosophical accounts of ontology (rather than the computer science accounts). In philosophy, ontology refers to the question of the meaning of being. This usually has three flavors. For instance in the statement "it *is* flooding" "is" (part of the verb to be) serves as a copula, connecting flooding and current events. A second question of being is an existential one, such as when you say "I am." And thirdly being can involve questions of identity, such as "this road *is* susceptible to traffic." Ontology investigates the meaning of all these statements.

So one way to think of the differences between "ontologies" and "ontology" is that the former asserts definitions and properties (known as substance ontology since the time of Aristotle), and the latter asks about their meaning and the nexus of relations in which it has its being. Unfortunately for us, computer scientists, and now GIScientists, chose the word "ontologies" unwisely, insofar as ontology means something very different in philosophy, and has done ever since the ancient Greeks introduced the term.

Here we can make a distinction first introduced by Heidegger in his book *Being and Time*. Heidegger's insight was to distinguish between ontical knowledge and ontological knowledge. Ontical knowledge is what GIScience has called ontologies; it is knowledge of the properties of entities and beings. Ontological knowledge however is "more primordial" and is concerned with being in general (Heidegger 1962: 31). The bandying around of the word "ontology" in GIScience has mistakenly lead some commentators to think they are asking ontological questions, when in fact they are working at the ontical level (Leszczynski 2009a).

So would the problem be solved if the word was changed? Unfortunately not. Importantly, in ontological accounts of being such as that put forward by Heidegger and the post-structuralist tradition which was inspired by him, you can never go from "ontologies" to being, as being is not just a question of adding up all the properties of something. Heidegger called this "the ontological difference." That is, ontology is not to do with essences, but rather existence – having a present, past, and futural possibility. (Heidegger once put it that "essence is existence".) A GIScientist can assert that a flood is a weather phenomenon but cannot answer the question of what it means to be a flood. This raises serious questions about the ability of "ontologies" (ontical knowledge) to say anything meaningful about lived experience and being. On this account, we don't learn anything about being in the world by abstractly staring at something and listing its properties. Rather, we need to *encounter* the world in its being. As the poet Rumi puts it "since in order to speak, you must first listen, learn to speak by listening."

Heidegger doesn't say that we don't need ontical enquiry (indeed it is the domain of science), but on the contrary we need to investigate it in conjunction with ontological enquiry in the widest sense. And what is ontology then? Roughly, ontological enquiry examines the horizon of possibilities for ways of being to occur (including of course the scientific way of understanding the properties of entities, i.e., ontical knowledge). He spends quite a bit of time on this in his book, which needn't detain us here. But it is an absolutely critical difference for GIScience to grasp, because purely enquiring about properties, without doing any ontology, will never uncover what kinds of ways of being are possible for us right now, or what kinds of worlds we live in.

In many ways ontology is a study of us as we are the only species (so far as we know) that has an understanding of being. This would require an account of our existence (an existential GIS? An "anthropological" GIS? A "worlding GIS"?), in order to investigate different *worlds* cross-culturally. How does being become intelligible?

Richard Polt (1999) offers the following example of why "ontologies" are insufficient to answer this question. Consider "Dr. P.," a patient of the famous neurologist Oliver Sacks. Dr. P. has a neurological disorder, such that when offered an everyday object and asked to describe it he says:

> "A continuous surface" he announced at last, "infolded upon itself. It appears to have" – he hesitated – "five outpouchings, if this is the word" . . . later, by accident, he got it on, and exclaimed, "My God, it's a glove!" (Sacks 1985: 14)

Dr. P.'s condition manifests itself such that he correctly sees the properties of an object, but has no idea what it is until it is put to use in a meaningful way. If he hadn't got the glove on, he could have listed the properties of it endlessly, without approaching what it *is*.

Compare computer and GIScience "ontologies" of semantic triples to our encounters with landscapes as described and evoked by artist Steven R. Holloway. Holloway, who "considers my work to be a gift from the place itself, essentially spiritual," works with maps, photographs, and performance art to speak of these landscapes. The illustration overleaf, for example (Figure 8.1) forms part of an ongoing series of works along Clark's Fork of the Columbia River near Missoula, Montana. What sensibility and encounter is evoked here? Holloway uses a "lensless" or pinhole camera in order to properly encounter the river. The slowness of the camera's operation ensures that no quick snapshot can be made, but rather a sustained dwelling. Using principles of generosity and awakening from Buddhism, but also the language of Heidegger (enduring, encountering phenomena, and care), Holloway writes:

> My work with lensless imaging began in the early 90s when I sought to change my photographic approach from taking thousands of 35mm images to taking just a few large format ones. . . . These lensless, or pinhole, cameras turned out to be a great success as they enabled me, as a geographer and map-maker, to begin to see and experience time and space as one interwoven experience . . . I take photographs in stopping to observe, not as an end object. In my practice of "right map making" I make repeated visits to a place-time event. In the course of these visits I count the birds, the stars, the flowers, the drops of rain. I count whatever I can. And when I drift from counting I measure. I measure the flow of the river, the air temperature, the water and the soil temperature, the dampness of the earth around me, the colour of things near and far, the weight of the air. I do everything possible to cease, to enter into the experience of the place-time. I respond from this experience of being with and within the events, in this case a bend in a river, using the languages of colour, shape, line, number, word and dimension. (Holloway, pers. comm.)

Figure 8.1   "River Bend View: Visit 14, 20 November 2004, 30 Seconds." Steven R. Holloway. Used with Permission.

How different a sense of place this is from listing the entities in the landscape with their properties!

Rumi says:

> Asleep on the bank of the river, lips parched,
> You dream you are running toward water.
> In the distance you see the water of your desire
> And, caught by your seeing, you run toward it . . .
> While he dreams of the pangs of thirst,
> The water is nearer than his jugular vein.
>   (Mathnawi IV, trans. K. Helminski)

Study of the meaning of being necessarily involves not only philosophy but other human sciences in that meaning is given to humans. In a recent effort to think through the ontology of mapping, Rob Kitchin and Martin Dodge argue that maps are not representational but "processual," that is, they emerge only through practices (Kitchin and Dodge 2007) and not through definitions. Obviously this has a lot in common with the point just made. They suggest that such an understanding can connect up mapping as an applied or technical practice with mapping understood as a form of power/knowledge. Neither of these is sufficient by itself, they argue (whether such a connection is possible in this manner remains to be seen). They use a biological term, ontogenetic (or ontogeny), which in that discipline refers to development of an organism, in a new way, to refer to the fact that they see being (*onto*) as being created or born (*genesis*). Thus for them, the being of maps is continually created afresh. This is very different from "ontologies" which explicitly try to tie definitions down and secure them.

Perhaps then, those pursuing "ontologies" in GIS and those pursuing philosophically driven "ontology" are two ships that pass in the night, or two tensions that are pulling across mapping (Figure 1.1). This is the gist of two papers by Leszczynski (2009a; 2009b), who argues that critical geographers have failed to distinguish between ontologies and epistemology. For Leszczynski, this failure has meant that critical cartography and GIS conflates mere discourse (epistemology) with the material conditions of GIS (the ontologies). Using a broadly critical realist approach, she pushes back against critical geographers who she says need to rectify this failure and pursue ontologies. Although there are certainly grounds for disagreement with this characterization (Crampton 2009a), it is certainly true that GIS (whether critical, Big, or traditional) has moved well beyond the GIS wars. It is nuanced and theoretically informed, it is increasingly diverse, but it is also not deflected from its traditional technical research agenda. As I observed in Chapter 1 then, the field of mapping continues to experience a whole series of competing tensions and it is likely that these will not be resolved any time soon.

Note

1 The original email is still available in the Usenet archives on Google: see http://tinyurl.com/623eb2.

Chapter 9

Geosurveillance and Spying
with Maps

"We are under surveillance all the time."
New York City Mayor Michael R. Bloomberg, October 2007

Fear of a BlackBerry Planet

On the fifth anniversary of 9/11, an unclaimed backpack containing a BlackBerry personal digital assistant was found on board United flight 351. According to news stories, the airplane was diverted to Dallas where it was searched, but nothing unusual was found.

> The flight crew on United flight 351 from Atlanta to San Francisco elected to stop in Dallas around 7 a.m., said SFO spokesman Mike McCarron, after finding a backpack on board that no one claimed. The unattended backpack contained a PDA device, McCarron said, "like a Blackberry." (Lagos 2006)

There was a time, perhaps not too long ago, when finding a BlackBerry would have meant "finders keepers, losers weepers." Now however, it is grounds for emergency action by the state.

This example is by no means unique. United Airlines flight 919 was diverted in 2004 when Yusuf Islam, the pop singer formerly known as Cat Stevens, was found to be on board. The US government refused to disclose what connection, if any, he had with terrorism and he was returned to the UK without charge. In May of 2005, Alitalia flight 618 was similarly diverted to Bangor, Maine, after a passenger's name was found to match one on the no-fly list. After being briefly detained, the passenger continued his flight, again without charge. (In October 2006, the CBS show *60 Minutes* revealed that the no-fly list was riddled with errors.) There are reports that a number of women have been forced to drink their own breast milk to demonstrate its safety (even if packed in the allowable three-ounce bottles), and in both August and September 2006 flights were diverted because the cabin crew found *bottles of water* on the plane (Bernhard 2007; King 2006; WSOCTV.com 2006).

All of these events raise the question of how the politics of fear provides a "rationale" for the use of geographical surveillance (geosurveillance). Fear, as an emotion, is receiving increased attention in the geographic literature. Known as "affect" (Lorimer 2008) we can consider how emotion – fear, in this instance – is instilled and reproduced by oneself, by others, or by institutions. To what extent do geospatial technologies help us understand affect? Are they equally likely to manifest at any given place, or are there, as we might expect, distinct geographies of affect that these technologies can help us understand (Kwan 2007)? How do these same geospatial technologies play a role in assisting or promoting societal fear.

In assessing this role we should be careful not to assert that geospatial technologies are *essentially* negative. As I discuss further in Chapter 12, we can think of geospatial technologies as a process of "creative destruction." Indeed, it is the heterogeneous and multiplicitous possibilities of mapping that give credence to recent efforts to "reconstruct" it (Schuurman and Kwan 2004), or to explore "counter-mappings" (Harris and Hazen 2006). Fear and possibility: can we steer a third pathway between these two positions, seeing geospatial technology and forms of government as "co-constructed" and as situated practices (Lyon 2003)?

Forms and History of Geosurveillance

Geosurveillance can be defined as the surveillance of geographical activities. "Surveillance" is a combination of the French words *sur* (over) and *veiller* (Latin *vigila*) to watch. (A related word is vigil.) The earliest attestation in English is from 1802 but the word possibly came to prominence during the Terror of the French Revolution (1793–4).

Because geosurveillance includes a wide range of activities, including not only surveillance of migration, travel, and movement, but also the distribution of people and things in territories or spaces, the range of geosurveillant techniques is also potentially wide. Two categories are nonetheless identifiable. Many accounts focus on the *individual*. Such geosurveillance includes tracking devices like passport RFIDs (radio frequency identification), cell phone geolocation, CCTVs (closed circuit televisions), avian influenza cases, organ donors, criminal offender monitoring, and so on. A second category of geosurveillance focuses on monitoring *groups and populations* as a whole. Here a population is not just a sum of individuals, it is an object of enquiry in itself. It has regularities, for example birth, death, and reproduction rates.

Surveillance is not new. The Bible describes a census in the Book of Numbers. However, extensive and systematic surveillance is more characteristic of modern societies. As Lyon (1994) discusses, older, pre-Foucauldian explanations of surveillance drew on economic and bureaucratic factors. In Marxism, surveillance was established with the advent of modern capitalism because it was a necessary part of managing people at work: "[h]ence what we now know as 'management' was developed to monitor workers and to ensure their compliance as a disciplined force" (Lyon 1994: 25). Max Weber extended the necessity of surveillance: "for him, surveillance is bound up with bureaucracy . . . [m]odern organizations are characterized above all by their *rationality*" (Lyon 1994: 25, original emphasis).

Lyon argues that systematic surveillance emerged with "the growth of military organization, industrial towns and cities, government administration, and the capitalistic business enterprise . . . it was, and is, a means of power" (Lyon 1994: 24). Surveillance (and geosurveillance) then are forms of knowledge – of knowing how much, where, and by whom – that are tied to forms of power.

During the First World War, new forms of mass (population level) surveillance were introduced, many of which had a geosurveillant component. In order to assess their suitability for a large-scale war, new recruits were subjected to batteries of tests and measurements (a practice known as anthropometry). Information was also recorded on other types of people such as war objectors, suspicious persons, and so on. Citizens were issued identity cards, state-issued photographic identification, and "securitized" passports. Interception and decryption of messages and transmissions was developed, and further extended in World War II, as we saw in Chapter 5.

In recent years, the increase of surveillance (particularly electronic surveillance) has given rise to the label "the surveillance society" (Lyon 1994; Pickles 1991) to capture the idea that surveillance has become institutionalized. As the computer age has progressed, more attention is being paid to digital or electronic surveillance and many people expect to be under surveillance as a natural state of affairs. According to a recent poll, 1 in 5 Americans (about 24 million households) think the government may have listened to their calls (CNN 2006); a number far above the probable reality, but reflective of a politics of fear.

A true surveillance society was postulated in Christopher Priest's 1978 short story "The Watched" in which confetti-sized surveillance devices called scintillas were spread around everywhere:

> The way of life [in the affected areas] had been permanently altered: scintillas were used in such profusion that nowhere was entirely free of them. They were in the streets, in the gardens, in the houses, in shops, offices, airports, doctors' surgeries, schools, private cars. You never knew for sure that a stranger was not listening to you, recording your words, watching your every action. Social behaviour changed: away from home people moved with neutral expressions, said or did nothing that was not bland or apparently harmless.
>
> At home, not because they assumed they were unwatched but simply because they were at home, they broke free and acted without restraint . . . wherever you went you were one of the watched. (Priest 1978/1999: 192–3)

Although it is difficult to say definitively, we may be the most-watched society in history. A commonly reported statistic, for example, is that the typical Londoner appears on CCTV more than 300 times a day. Many surveillant technologies have been increasingly deployed in the pursuit of security since 9/11 and the passage of laws such as the USA PATRIOT Act of 2001 (HR 3162). While many citizens of the United States, Canada, and other countries have been generally willing to accept the need for surveillant measures abroad or of foreigners in return for security (a bargain that is itself doubtful), they have generally drawn the line at domestic surveillance. Over the past few years however, it has gradually become clear that it is not easy to draw a bright line between foreign and domestic surveillance.

The Panopticon

Priest's speculations about a possible society of total surveillance draw some support from those occasions when societies have put such a system into effect. In East Germany for example, the secret police or Stasi employed an estimated 300,000 German citizens as informants. Coupled with a repressive authoritarian state, GDR residents did expect to be under surveillance most of the time and censored themselves accordingly. Although there were those who spoke out or resisted, they would be subject to increased surveillance.

A powerful dramatization of this is given in the award-winning film *The Lives of Others* (2006). Set in East Germany before the fall of the Berlin Wall, the film is the story of two men; the writer Georg Dreyman (Sebastian Koch) and Stasi member Ulrich Mühe (Hauptmann Gerd Wiesler) who is assigned to watch him. At first Mühe performs his job admirably, but as he continues to watch, he develops sympathy for Dreyman. At the critical time he disobeys the Stasi and is demoted to an operator of the machine that steams open letters (not a prop in the film, but one of the actual machines; it could open 600 letters per hour). So Dreyman survives, and in a final scene after the end of the regime, we see Mühe (now a postman) pop into a bookstore to buy Dreyman's latest book which he poignantly finds is dedicated to him. Was he forgiven? The last line of the film "Es ist für mich" (it's for me) finally and powerfully breaks out of the "we-society" that everyone had been living in. There was, despite tremendous damage to everyone's lives, finally room for the true individual.

Few societies can match the blanket surveillance achieved in East Germany, but many other societies, even those in the West, quite frequently achieve near-total surveillance on parts of their citizenry or in particular geographical settings. In the United States surveillance can be carried out deploying another kind of fear; not of the government itself, but of an external threat. Thus a government need not be a totalitarian one in order to successfully mount comprehensive surveillance, indeed it can describe itself as acting in the best interests of its citizens.

In this respect one of the most famous examples of surveillance is the ideal prison known as the "panopticon" (all-seeing) which was proposed by the social reformer

Jeremy Bentham in the late eighteenth century (Bentham 1995). Although people sometimes assert that Bentham's plans were never realized in practice (Monmonier 2002b), in fact some 300 prisons around the world were built on "panoptic" attributes. In Bentham's ideal prison, the wings of the cell-blocks were arranged as spokes from a central hub. The guards, who would be located in the central hub, could observe (directly or with the use of mirrors) all the cell-blocks without moving. Furthermore, the guards were hidden from the prisoners' view, and so could observe without being observed. This aspect has had a very strong metaphorical grip on the idea of surveillance.

Bentham's ideas were introduced to a fresh audience in Foucault's work on the history of the prison. His book *Discipline and Punish* (*Surveiller et punir* in French) famously gives an account of surveillance of the individual (Foucault 1977). Foucault argued that we as individuals are situated within a myriad of power relations, which he called "discipline." This was the origin of the idea that we live in a "disciplinary society," a phrase he first used in 1974 (Foucault 2000d), and again in *Discipline and Punish* the following year.

For example, Foucault discusses the Eastern State Penitentiary on Fairmount Avenue in Philadelphia, which has panoptic features, in particular the central rotunda from which guards could see down the spokes or corridors. The penitentiary's primary goal, derived from liberal Quaker principles, was prisoner isolation in order to achieve "penitence" (hence penitentiary). The Quakers thought that by essentially isolating each prisoner for up to 23 hours per day, the prisoner would be driven to seek within himself the attitude of sorrow and admission of guilt. Only then could the prisoner repent and be cured. In the hands of the Quakers the imprisonment was not so much for punishment as it was for their own good. Still extant, the penitentiary was constructed in 1829 and operated until as recently as 1971 (Johnston 1994; Teeters 1957). See Figure 9.1.

The goal at the penitentiary was complete isolation; the guards even wore slippers over their shoes as they patrolled the cells so that no prisoner might know when they passed by.

This argument, that the practices of surveillance are established for the benefit of those being surveilled is a characteristic discourse of modern, neoliberal societies. If for the authoritarian society surveillance was a means of social control through fear, liberal societies such as the UK and USA employ surveillance through another sort of fear; that is, fear of the outsider who poses a threat or risk.

Surveillance, Risk and the "Avalanche of Numbers"

We can follow Foucault's argument for a little bit to see why this might be. The fact that "risk" is such a big part of our society is no happenstance. For Foucault, risk-based societies emerged hand in hand with new forms of juridical governance. Prior to the legal reforms of the eighteenth and early nineteenth centuries, the law

Figure 9.1   A view down one of the "spokes" or corridors in the Eastern State Penitentiary, Philadelphia. Source: photograph by author, 2006.

focused on the nature of the crime committed, the evidence of guilt or innocence, and the system of penalties that would be applied. In other words: crime and punishment. The person of the criminal was important only insofar as he or she was the individual to which the crime would be attributed. With the reforms, this hierarchy was reversed, the crime was merely an indicator of something more significant; the "dangerous individual" (Foucault 2000a). The law was now interested in the potential danger of the individual: "the idea of *dangerousness* meant that the individual must be considered by society at the level of his potentialities, and not at the level of his actions; not at the level of the actual violations of an actual law, but *at the level of the behavioral potentialities they represented*" (Foucault 2000d: 57, original emphasis). Thus discipline had to be appropriately tailored to the perceived threat of the individual.

But for Foucault discipline was only one of a number of forms that power relations could take. If discipline describes power at the level of the body, then a later concept, "biopolitics" describes power at the level of the population (Foucault 2000b). And as we shall see, while individual tracking is important and news-worthy, biopolitical population geosurveillance is far more extensive in today's risk-based society.

For example, just prior to the National Republican Convention in September 2008, police and federal officers performed pre-emptive raids on at least six houses (Greenwald 2008a). Although the groups concerned had performed no illegal actions, they were aided by informants placed within the groups to observe behavior, according to newspaper reports (McAuliffe and Simons 2008). Lawyers for those affected noted that the raids were "purely anticipatory" in nature and that by carrying automatic weapons the SWAT teams are designed to quell dissident political protest (Greenwald 2008b). As we noted earlier, the bright line between foreign and domestic surveillance seems increasingly hard to maintain.

The Italian philosopher Giorgio Agamben put it succinctly:

> [President] Bush is attempting to produce a situation in which the emergency becomes the rule, and the very distinction between peace and war (and between foreign and civil war) becomes impossible. (Agamben 2005: 22)

In this light it is possible to see that today's most important geosurveillant resources are *biopolitical* technologies that include not only the well-known border fences and motion detectors (border biopolitics, see Amoore 2006), but a new generation of mapping technologies. The kind of security desired by the United States and other countries depends on a whole suite of digital spatial mapping and so-called "locative" technologies. These locative technologies allow people and objects to be geosurveilled, that is, to be tracked, marked, noticed, and logged as they move from one place to another. It is not a question of identifying which areas are at risk therefore, because everything is at risk to different degrees. Geosurveillance is required to be equally co-extensive with that risk, that is, everywhere. Blanket geosurveillance is therefore a logical outcome of the state's representation of its residents as risk factors who need to be controlled, modified, and logged.

When this discourse is successfully deployed, people are usually quite willing to trade off, as they see it, freedoms for security. Therefore it is not surprising to find that governments interested in surveillance employ a whole range of techniques to instill fear, not of themselves, but of the stranger, the immigrant, the foreigner, or any group that does not fit received *norms*. For example, in a poll conducted by *Newsweek* in July 2007, Americans were asked if the FBI should secretly wiretap mosques to "keep an eye out for radical preaching by Muslim clerics." Over half the respondents (52 percent) agreed (*Newsweek* 2007).

How did we get to such a situation?

There are several steps involved. First, we have to understand norms in a calculative, statistical manner. With the emergence of demography and statistics as sciences in their own right in the nineteenth century, demographers, geographers, urban planners, and political economists such as Charles Dupin began to observe and quantitatively investigate things like birth and death rates, hygiene, accidents, age of marriage, number of children, education, divorce rates, as well as anomalies. These statistics were critical as the etymology of the term indicates. It was derived in the late eighteenth century from the German *Statistik* meaning "state-istics," statistics for/of the state (Hacking 1990; Oxford English Dictionary 1989; Shaw and Miles 1979). As we saw in Chapter 6, by the mid-nineteenth century, many of the modern forms of statistical thematic mapping used in today's GIS were in place.

Second, after the Belgian statistician Adolphe Quetelet developed the idea of *l'homme moyen* (average man) in 1835, then criminologists, for example, were able to assess and map out the areas where crime was above or below normal. Knowing this, they could issue policy prescriptions.

Third, and underpinning all this, the state had to collect whole reams of new knowledge. In the nineteenth century then we see a tremendous rise in the amount of information being collected, what Ian Hacking once called the "avalanche of numbers" (Hacking 1982).

It is therefore not incidental that in the early nineteenth century the major forms of mapping concentrated on showing exactly these factors; where pockets of disease might lie, and where trade went. To take just one example, an innovative series of proportional flow maps were made by Joseph Minard showing wine exports from France (Friendly 2002; Robinson 1967; Wainer 2003). These were among the first uses of proportional symbols on maps. Or again, take the famous case of John Snow, the father of epidemiology or bio-surveillance, who, if he did not use the map to discover the source of cholera in London, certainly used its rhetorical powers (Johnson 2006). Much of this data was derived from censuses – much improved and professionalized in America during the second half of the nineteenth century as we saw in Chapter 6.

Geosurveillant Technologies and the Biopolitics of Fear

How and why does a climate of fear, and more specifically, a biopolitics of fear operate? And how might geosurveillant technologies be involved? Drawing a little bit more on the work of Foucault, in this section I suggest that it requires at least three critical practices: *divisions*; *geosurveillant technologies*; and the *risk-based society*.

Divisions

The first step in activating a politics of fear is to create and constantly reproduce a whole array of divisions between "us" and "them" (e.g., normals vs. abnormals; insiders vs. outsiders). Those who promote these divisions (whether it be governments or local actors) wish to establish certain kinds of power–knowledge relations with the "others" in question. Foucault called these "dividing practices":

> The subject is either divided inside himself or divided from others. This process objectivizes him. Examples are the mad and the sane, the sick and the healthy, the criminals and the "good boys." (Foucault 1983: 208)

For example, immigrants have often been the target of surveillance, not necessarily because of their economic potential but because they represented a threat. In the early twentieth century immigrants were subjected to a series of race-based quotas in American immigration laws (Crampton 2007b). Since the quotas were derived from the census, it becomes apparent that censuses are not just collections of data, but a political technology of population management. This is a major difference between Foucault's work on surveillance and that of Marx and Weber. For Foucault, surveillance is neither about class relations nor bureaucracy, but about the biopolitics of "the essentially aleatory events that occur within a population" (Foucault 1983: 208). Here aleatory means that people are, essentially, free – what one person does is a matter of chance – but also that there are contingent regularities that could be modeled by a theory of probability. Given birth rates in an area, one could predict if a new school would be needed, although not whether any specific couple would have children. One could now talk about "group norms" and "deviations" from these norms. These norms were used in constructing a series of dividing practices. As Hannah has observed, this biopolitics of subjectivity reveals how spatial partitions such as borders and territories serve to regulate and govern populations in "mappable landscapes of expectation" (Hannah 2006: 629).

These concerns translated into concrete proposals for town planning as Margo Huxley has shown (Huxley 2006). Towns and cities had to cope with movements of people (circulation) due to trade coming and going into the city, but also, in a time when cities no longer had closed walls, the threat of vagrants and thieves coming in from the country. Furthermore, to ensure the general health of the residents, and to prevent insurrections, good kinds of circulation had to be promoted while diffusing or preventing bad kinds (for example, by building wide boulevards which the mob could not blockade) such as Haussmann's Paris (Harvey 2003).

Geosurveillant technologies

Second, in order for these divisions to work it is necessary to deploy a variety of technologies of surveillance, spatial tracking systems, and geosurveillance. The purpose of these technologies is to collect, sort, manage, and display spatial information about the identified groups, often with a view to managing or overseeing their geographical distribution or movement. These technologies run the gamut from long-standing data collection efforts such as the census, to newer efforts such as the American government's use of warrantless wiretaps, mechanisms of movement control (e.g., border security), and the cartographies of these landscapes.

For example, take crime mapping which works through "geoprofiling." Geoprofiling is a technique for determining the typical spatial patterns of an individual with the goal of predicting that person's behavior or targeting them for surveillance. With geoprofiling maps can easily be made of crime hotspots and coldspots.

Crime maps enable geoprofiling to isolate behavior that does not conform to the norm. But profiling can be controversial. After a series of high profile incidents on the New Jersey turnpike in which African American drivers were disproportionately stopped by the highway patrol, it was charged that the police were stopping blacks because of who they were, not because of their actual behavior (Colb 2001). That is, criminality judgments were made on the basis of *potential dangerousness*, rather than actual offenses being committed (i.e., the searches were made without probable cause).

Consider finally that a significant amount of GIS research is performed in risk analyses. Our inability to satisfactorily assess risk notwithstanding, the discipline seems wedded to a risk-based approach (Cutter et al. 2003). Obviously, only some small subset of this research is concerned with assessing risky populations for purposes of national security (a recent search of the main ISI database yielded over 1,000 research articles on "GIS and risk"). But here it is worth reminding ourselves that although they are often portrayed that way, technology and politics are not two separable domains. In particular, there are long-standing ties between mapping, GIS, and military and intelligence agencies, as was discussed in Chapter 5 (Cloud 2002). For example, in the USA one of the leading GIS companies is a "strategic partner" of the NGA (National Geospatial-Intelligence Agency). There are now major gatherings of geospatial intelligence personnel at an annual conference known as GEOINT (geographic intelligence).

In light of this, we need to understand how mapping and other sources of geographical knowledge act to produce the politics of fear. The answer is not to cease using GIS and mapping technologies (or only to use the "good ones" – whatever they are), but rather to be careful and critical about the knowledge that is constructed with them and the subsequent political rationalities that are supported by them. If the profession is to establish an ethics of practice as some desire, then it at least also ought to have ethical and legal protections for those who refuse their labor (although how enforceable such a provision would be is unknown).

The risk-based society

Once data have been collected through surveillance, they have to be rationally assessed. Today, this is carried out through the model of risk. There are several elements to this.

First, the divisions previously established are used to sort out data into categories. Second, each category has an associated degree of risk. Third, all members of that group are assumed to pose the same degree of risk. If you belong to a high-risk group then you are also a high risk, whatever your individual qualities may be. Using risk (or its cognates, threat and security) shifts the judicial process from one of prosecuting offenders *after* the crime (a question of individuals) to anticipating and pre-empting actions by those within high-risk groups (a question of populations).

What is wrong with all that? There are a number of reasons why adopting a risk-based approach for human affairs is undesirable. First, in the shift from offender prosecution to profiling there is a danger of wrongly identifying individuals based on group membership (known technically as a false positive). A false positive is a false indication of a positive finding. It arises when average group-level characteristics are used to derive information about the individual. A common example occurs in the context of automobile insurance. Many insurance companies ask you for the zip code where the vehicle is owned. Using accident data for this zip code they then assess the premium for that vehicle. Two people with clean driving records will pay different premiums based on where they live – not their personal records. (Some companies do make small reductions in the premium for such records.) This philosophy of imputing individual characteristics based on group membership lies at the heart of profiling, stereotyping, and racism and has therefore often been rejected (for example in "random" car stops by the police).

In some contexts these false positives are acceptable, but when dealing with human subjects a false hit can be devastating. A false positive can lead to arrest, detention, or deportation. Where base rates of the activity are low (as is presumably the case with terrorism in the West) false hits can far outnumber real hits. Since it is not determinable prima facie whether a hit is truly or falsely positive, all such hits have to be investigated (draining resources). It is therefore not surprising that most terrorist-related arrests do not lead to prosecutions. In the United Kingdom fewer than *4 percent* of the people arrested under anti-terrorism laws have been convicted of terrorism (Morris 2007).

Second, in order to make these risk assessments extensive group-level information (the zip code in the example above) must be collected. While the state would appear to gain from the efficiencies of profiling, its citizens pay the price of more extensive surveillance. For example, in 2005 the United Kingdom announced that it would be the first country to track every journey by car (Connor 2005). Every day automatic cameras will record car license plates 35 million times, capturing time, date, and location with GPS. In these data-mining situations it is not only ironic but significant that the vast majority of the data must of necessity be of innocent citizens.

A third problem with risk-based analyses is that our evaluations of risk are extremely poor. When we are faced with making a judgment in a situation with probabilities rather than certainties (which is the case with risk) most of these judgments demonstrate consistent biases or errors. Yet at the same time most people believe that their judgments are accurate and superior to those of other people. In one study in 2000 for example, in the context of the estate or death tax benefiting only the top 1 percent of earners, fully 39 percent of respondents thought that this would benefit them. These sorts of findings are well known, and were first extensively demonstrated in the early 1970s, most famously by Tversky and Kahneman (Tversky and Kahneman 1974).

These errors in judgments are not because people are stupid, but because they are affected by the context in which they are made. Recent work on risk assessment for example has uncovered the so-called "dread risk" or low-probability, high-consequence risks (Gigerenzer 2004). A dread risk is often incorrectly expected in a climate of fear, and can seem reasonable. But it will usually lead to disastrous consequences. Following the 9/11 airplane attacks on the Twin Towers for example, many people feared another airborne attack and elected to drive to their destinations.

But since driving is so much more dangerous than flying, it is estimated that this surge in driving led to the additional road deaths of 1,500 people in the following year (Gigerenzer 2006).

Yet while fear becomes more pervasive throughout the USA, statistically the country has never been safer. Americans live longer, healthier lives (some 60 percent longer in 2000 than 1900), have better access to clean water and food, and enjoy safer workplaces. There is an increasing mismatch between perceived danger and actual risk, much of it caused by the media. Siegel cites the fact that between 1990 and 1998 "the murder rates decreased by 20 percent, while murder stories on media newscasts increased by 600 percent (not even counting O. J. Simpson)" (Siegel 2005: 56–7).

"Map or Be Mapped": Counter-Mapping, Spaces of Resistance, and the Ethics of Forgetting

So where does this leave us? One thing is clear, geographers can do much more to discuss the deployment and involvement of mapping technologies in surveillance. According to a recent study of the literature, only 4 percent of articles in leading GIScience journals are about the social or theoretical aspects of mapping technologies (Schuurman and Kwan 2004).

As one GIS expert recently put it however, mapping technologies are not just techniques, but "entire new ways of seeing" that weave together the technology, the political and the social (Klinkenberg 2007: 357). If more GIS leaders were to express such views the result would be decreased attention to hyped-up and unlikely dread risks and the politics of fear.

In lieu of that moment however, it might be worth recalling that one component of critical cartography and GIS is that it is oppositional. Earlier in the book, we investigated ways in which GIS and cartography could operate "after critique." In the last part of this chapter I'd like to briefly indicate some further spaces of resistance.

We might begin by looking at bit more closely at what the term resistance means, a term that has had its fair share of romance (Sparke 2008). Autonomous actions and "fighting against the system" are attractive, but are they just too idealistic? It's useful to begin an answer to this question by rejecting a zero-sum game; that you've achieved nothing if you haven't achieved a pure revolution. As we saw with Elwood's ethnographies in Chicago, the more typical outcome is neither cooptation nor resistance, but some mix of the two. Taking this further, in her recent book on children's everyday lives, Cindi Katz has troubled the notion of pure resistance by supplementing it with some other "r-words," that is, reworking and resilience (Katz 2004). (A fourth and less welcome possibility, revanchism, occurs when vengeful policies such as welfare "reform" are put in place.)

For Katz, reworking comes into play when the terms of the game are changed a little, without necessarily resulting in the end of all inequalities in the relationship.

Projects of reworking include both redirecting available resources and retooling yourself as a different kind of actor (e.g., by becoming a more political person).

An analogous tactic is described by Foucault in one of his last lecture series, on the topic of frank speech (Foucault and Pearson 2001). Foucault tells the story of the philosopher Diogenes confronting Alexander the Great. In this story Diogenes is the model of the frank speaker (in Greek known as *parrhesia*), and Alexander the supremely powerful ruler. Despite this imbalance Diogenes has something that Alexander does not; besides this frank speech he is unafraid wherever he goes and lives life the way he wants despite having no money. And Diogenes begins by insulting Alexander, telling him that real kings don't go around heavily armed, as if trying to prove something. Following this insult we are told that Alexander came near to throwing his spear, but that Diogenes, instead of pleading for his life, asserted instead that Alexander could indeed very well kill him, but then he wouldn't hear the real truth. So Alexander pauses for a moment and considers that Diogenes may be saying uncomfortable things, but he does say the truth and speak frankly. So Alexander agrees not to kill him and agrees to Diogenes' new "contract": either kill me or hear the truth. So now the stakes of the game are changed, a new limit is set and Alexander cannot kill Diogenes if he wants to hear the truth. In this example then, we could say that Diogenes "reworks" the relations between himself and the powerful authority, to his own advantage.

Resilience, as the name implies, is those efforts at autonomous initiative and recuperation. All three may overlap or it may be possible to build up from resilience through reworking to full-on resistance. All three tactics are likely to involve some measure of risk. As rural landscapes change, for example through deforestation, local people may find that their communal knowledge and way of life is threatened, unless they can perform some act of resilience (in Katz's example, the villagers underwent a marked expansion of their social territory). Such an expansion may not be possible of course, or may in turn generate new conflicts with neighboring communities.

In recent articles on indigenous mapping, the geographers Joe Bryan and Joel Wainwright have argued that it is not the map or the act of mapping per se that is critical to resilience. Neither the map nor the law after all actually *is* justice. In two cases they examined, the Maya of Belize and the Mayangna community in Nicaragua, the outcome was new legal precedents and maps in support of indigenous land-claims. In the latter case for example, maps were used to contest a government concession to a Korean-financed logging company on lands claimed by the local community. After many years and appeals the court did find in favor of the community, at least partly on the basis of a map they had produced. In the Belize case, the Mayan communities in the south produced an atlas of all the lands the Maya have historically used; the *Maya Atlas* (Toledo Maya Cultural Council and Toledo Alcaldes Association 1997) and also filed a lawsuit against the state. Again, courts ruled in their favor, but realizing the gains has proved problematic.

Despite these legal successes, Bryan and Wainwright, who were involved in the map-making process in both cases, caution that it is not so much a case of over-turning the colonial legacy as reworking it. In other words, it's a larger and more complex issue than the "map or be mapped" strategy that we mentioned earlier of replacing bad colonialist maps with good anti-colonialists or indigenous maps.

Why? As they observe "[l]ike the law, maps are not instruments for settling indigenous claims. They are textual practices that weave together power and social relations. The effective indigenous 'counter-map', then, is one that unsettles the very categories that constitute the intelligibility of modern power relations" (Wainwright and Bryan 2009: 170). As they also write "[w]e contend that these new maps of indigenous lands are neither inherently good nor beyond question: they are open to multiple readings, and they may have potentially undesirable outcomes" (2009: 154). The emphasis for these researchers then, as so often in this book, is not on technologies per se, but the larger socio-political ramifications.[1]

One of the original forms of resistance to surveillance was mounted by the "Surveillance Camera Players" (SCP). They certainly demonstrate both reworking and resilience. This New York City-based group, who describe themselves as inspired by the Situationist movement, was founded in 1996. Their protests take the form of short, specially adapted plays which they perform in front of surveillance cameras. The group also publishes maps of known surveillance cameras, which they have scouted out. Their website and associated book (Surveillance Camera Players 2006). For example, in one play, *It's OK, Officer* (2002), they hold up a series of placards in front of a surveillance camera, which proclaim:

It's OK, Officer
Just Going to Work
Just Getting Something to Eat
Going Home Now . . .

And so on. Each placard is illustrated with a little stick figure drawing of the person walking, eating, or shopping in accordance with the words. The SCP have been monitored by the NYPD as they found out in 2007 when they obtained records about them which they had requested through the Freedom of Information Act.

From a mapping perspective, one of the most revealing activities are their maps of surveillance cameras in neighborhoods across the city. Over time, these maps have become ever more populated with cameras, until now it is impossible to say how many are in New York (one number mentioned is 15,000 cameras in public places in Manhattan alone by 2006). In the practical Situationist spirit in which they operate they have provided instructions on how best to map surveillance cameras (Figure 9.2).

My final example of "r-words" in this chapter draws from a proposal made by Martin Dodge and Rob Kitchin which they dub an "ethics of forgetting" (Dodge and Kitchin 2007). After noting that modern pervasive computing is reaching a state of near-total observation (if not yet at the level of "The Watched"), with computers deeply embedded in our daily lives, recording our every move, they suggest that in response we need to see forgetting as a positive thing. For example, some companies

Figure 9.2   Surveillance Camera Players performing in front of public surveillance
camera. Source: Surveillance Camera Players (2006).

are now working on wearable computers that will record you (audiovisually) as you
perform your daily routine, and that with storage space so cheaply available this
can all be logged: the "life-log." (How much is a life worth: *x* gigabytes?) If this
smacks of digital enslavement (geoslavery in Dobson's memorable term [Dobson
2006]), then the only response should be to *resist* by advocating forgetting (or
digital erasure). Dodge and Kitchin ask, consider what happens to a person's life-
log at death? Or what about a child's life-log; should it be accessible by others?
Will life-logging actually be voluntary, or, given its likely monetization possibilities,
will you have to "play the game" to stay in the system (refuse and you find the
grocery store cannot serve you)? What about editing these life-logs; as people become
more dependent upon them, will they be subject to editing what are essentially your
memories (shades of the film *Eternal Sunshine of the Spotless Mind*)? Maybe these
questions are not yet answerable, but they are rapidly coming into sight as we depend
ever more on technology. In writing this book for example, I relied heavily on
technology such as emails where I stored websites or books I needed to follow up
(I send emails to myself), global keyword searches across chapters to make sure I

didn't repeat myself, and at one point after my hard drive crashed with the only copy of my text on it, an online data recovery tool that was able to piece together – correctly I hope! – about 90 percent of the draft.

In Dodge and Kitchin's idea, the recorded data about you would be subject to loss and error. Parts of it might fade away over time for example, so all those license plates being recorded in the UK would become incomplete. Or some of the data may be masked or made less precise and more fuzzy. In other words, digital memory would become more like human memory. There is something to this I think, if you consider confidentiality laws and practices in public health, such as HIPAA, or IRB procedures (especially informed consent). In a geographical study of disease for example, individuals cannot be identified by location, but to enable the analysis to be performed can be fuzzed up through "spatial masking" (Kwan and Schuurman 2004). The data are still extremely useful, but do not betray confidences. This is not to say that IRB protocols (developed in the context of research abuses in the 1940s and 1960s, the utility of which is often questioned in other fields such as geography) are not themselves problematic.

In the next chapter we shall investigate in more detail perhaps the single most notable domain of surveillance, that of cyberspace or digital reality. Cyberspace – a term invented by the science fiction writer William Gibson – has been a staple landscape of science fiction since at least the late 1970s when Vernor Vinge wrote his famous novella *True Names* (Vinge 2001). "Consensual reality," whether imagined as a place you can enter, as in the classic movie Tron (1982), or today's hybrid virtual–physical modes of communication such as Twitter, blogs, YouTube, Facebook, email, and so on has distinct geographies that can be mapped. Information 24/7/365: does this portend the "death of distance"?

Note

1   In early 2009, an indigenous mapping project in Oaxaca, Mexico, ran into just these kinds of unintended consequences. The controversy surrounding this work, which was performed by geographers from the University of Kansas along with indigenous peoples, was discussed at the AAG Annual Meetings in Las Vegas. Although I refer to it again in the last chapter, these events are still ongoing as this book is finalized.

Chapter 10

Cyberspace and Virtual Worlds

Science Fiction?

Google Earth (GE), Microsoft Virtual Earth (VE), and NASA's Word Wind exploit the idea of the "digital earth" (see Chapter 3) popularized in the 1990s by then Vice-President Al Gore. Imagine the following scenario, he said, which he acknowledged sounded a bit like "science fiction":

> [i]magine, for example, a young child going to a Digital Earth exhibit at a local museum. After donning a head-mounted display, she sees Earth as it appears from space. Using a data glove, she zooms in, using higher and higher levels of resolution, to see continents, then regions, countries, cities, and finally individual houses, trees, and other natural and man made objects. Having found an area of the planet she is interested in exploring, she takes the equivalent of a "magic carpet ride" through a 3-D visualization of the terrain. Of course, terrain is only one of the many kinds of data with which she can interact . . . she is able to request information on land cover, distribution of plant and animal species, real-time weather, roads, political boundaries, and population. She can also visualize the environmental information that she and other students all over the world have collected . . .
>
> She is not limited to moving through space, but can also travel through time . . . she moves backward in time to learn about French history, perusing digitized maps overlaid on the surface of the Digital Earth, newsreel footage, oral history, newspapers and other primary sources. (Gore 1998)

Gore's vision was not correct in detail (no head-mounted display or voice-control has ever become popular, nor do we need to go to a special museum to use a digital earth), but he captured several important points:

1. data are displayed "naturistically" as if on a planet as seen from space;
2. the display is interactive, allowing zooming and rotation (the "magic carpet ride," still an unfamiliar concept for geographical data in 1998) and querying by simple clicking on objects;
3. data from different sources can be integrated and easily layered;
4. time can be incorporated.

We saw in Chapter 3 how "peasants" (that is, everyday people such as students, amateur photographers, and so on) are adopting easy-to-use and widely available mapping tools. Importantly, not only are they using these tools, but they are providing stories about their lived experience. A facility like Google "MyMaps" allows people to annotate Google maps with their own content (text, photographs, videos) from their own lives. This content may not be meaningful on a wide scale (e.g., a stranger's wedding pictures), but for the people involved and among whom it is shared it is vital. After MyMaps was released in early 2007, millions of personal maps were created with it. Even Google was amazed at its popularity. It was not so much the use of their tools, but the creation and sharing of stories without the mediation of experts that was amazing. This is completely different from the traditional picture in geography and GIS which has always operated through a top-down, expert-driven process.

In effect: *Google Maps has become the Wikipedia of the geoweb*. It acquires and has data submitted to it, it edits and quality controls that data, and it publishes and makes that data modifiable.

This raises the critical question: is Google good for geography? (Here, I use Google as a shorthand for the geoweb more generally.)

Tensions in the Web: GIS vs. the Geoweb

Elsewhere in the book we have examined the possibilities afforded by the geoweb, and in particular the challenge that it makes to traditional Big GIS. But we have also noted that as a technology, it is not essentialist. As such "actually existing" technologies need to be understood in their particular socio-political contexts.

Some of the ways in which Google/the geoweb have come under criticism include: loss of privacy (e.g., through StreetView), issues of censorship (Boulton 2009); homogenization of maps (Wallace 2009); dumbing down of mapping – making maps that are very basic or lack richness (BBC 2008); contributing to the end of paper maps and/or the destruction of the mapping industry; and the proliferation of amateur(ish) maps made by non-experts ("McMaps," see Dodge and Perkins 2008).

Dodge and Perkins examine an apparent decline in map production and map use in some surprising places – including geography journals. At the same time the map is still emblematic of geography to the non-specialist: "[o]n the street and in the pub, British geography is still about maps" (Dodge and Perkins 2008: 1272). If corporate quasi-monopolies such as Google and Yahoo, which after all are media companies rather than cartography suppliers, have led to a proliferation of McMaps, are we any better off?

Recall Figure 1.1 and the tensions in cartography. Concerns over "undisciplined" and amateurish mapping are exactly what fuels the push to certify knowledge and establish bodies of knowledge. Our relationship to mapping is a profoundly ambivalent one (the unseemly perversity of maps that work too well).

If the geoweb is to grow up and be understood not just as the amateur version of what the professionals do, it needs to fight for legitimate recognition of its own professionalism. It should continue to critique and undermine the expensive, limited, and ill-designed capabilities of traditional GIS. It should not be happy to be just accepted as a minor co-player "alongside" professional GIS.

How can it do this? I would suggest by pointing to the following inherent factors which provide tremendous advantages for the geoweb:

1. "crowdsourced" data as for example in Wikipedia;
2. open source tools and services;
3. participation and syndication (the Web as platform).

Crowdsource

Here I will focus on crowdsource and the Web as platform (see Chapter 3 for discussion of open source). Crowdsource refers to the way that large numbers of distributed people can work on the same project in a very powerful manner, creating something where the whole is more than the sum of the parts. The online encyclopedia Wikipedia is a paramount example. Indeed, Wikipedia is an outgrowth of another encyclopedia project, called Nupedia, which accepted articles in a highly peer reviewed manner. Submission of articles was slow and the project eventually was canceled. Wikipedia on the other hand has an open, community-based approach where in principle anybody with a computer can edit or submit content (although material is subject to editing by power users and others at Wikipedia, but there is no peer review, nor is submission limited to "experts"). The key here is that Wikipedia and other projects are done by community consensus, not chaotically. These projects are self-organizing. The results are pretty clear: Wikipedia receives more than 450 times as many daily visitors as does the online *Encyclopedia Britannica*.[1]

The term crowdsource may be new but the principle is old. The Wikipedia entry on crowdsource cites the prize by the British government for the solution to the longitude problem in the eighteenth century and its eventual winner, John Harrison, as an early example of crowdsource.

When the skilled aviator Steve Fossett went missing in September 2007, Richard Branson, one of Fossett's friends, coordinated with Google to examine its imagery, and Amazon's crowdsourcing technology known as the Mechanical Turk was once again used as it had been in the case of Jim Gray. The Fossett search reportedly covered over 300,000 squares miles by some 50,000 people. (The search was unsuccessful, and some participants and search and rescue members have criticized it for distracting from official efforts since each flagged image has to be double-checked.)

While crowdsourcing has often been successful elsewhere, the lesson from the Fossett search indicates the parameters of the search by amateurs need to be clearly specified. The adage "many hands make light work" is central to crowdsourcing and volunteered geographic information (VGI) (Goodchild 2007), but it works better where the group's decision-making can be well coordinated. Surowiecki discusses methods of improving the wisdom of crowds, including these four principles:

1. have a diversity of opinion within the group;
2. have independence so that people's decisions aren't influenced by those around them;
3. have decentralization so that people are able to draw on local knowledge;
4. have good methods of aggregating opinions into a collective decision (Surowiecki 2004).

The Web as platform

The very nature of the Internet permits people not only to gain information but to *participate* as well. The Internet allows us to create new content and new knowledge. This participatory nature of information (sometimes called the "read/write" Web) includes community-based websites such as YouTube, Wikipedia, MySpace, Facebook, and the millions of blogs that constitute the "blogosphere" (the sum total of all blogs). Some of these blogs are read by a few people, some by thousands a day. Some make no money, some make over a million dollars a year (e.g., the tech blog Boing-Boing, see Tozzi 2007). Yet they all evince a desire and a need to share their views and ideas with others by using publishing tools that are widely and publicly available. It has been suggested on more than one occasion that the read/write Web may directly affect the kind of society that we live in and its politics – effectively, it has the potential to renew participatory democracy. Al Gore wrote recently:

> The Internet has the potential to revitalize the role played by the people in our constitutional framework. Just as the printing press led to the appearance of a new set of possibilities for democracy, beginning five hundred years ago – and just as the emergence of electronic broadcasting reshaped those possibilities, beginning in the first quarter of the twentieth century – the Internet is presenting us with new possibilities to reestablish a healthy functioning of self-government. (Gore 2007: 259–60)

The phenomenon of blogs is certainly one that bears watching. There are untold millions of blogs (Technorati gave up counting at 112 million), and most traditional media outlets now include blogs. Yet a word of warning: blogs also suffer from the "long tail" effect, that is, only a few are widely read (the head), while there are untold numbers that make up the long tail that are hardly read. In the Internet then, information and knowledge are just as unevenly distributed as in the physical world – the digital divide has certainly not gone away (Chakraborty and Bosman 2005; Zook and Graham 2007a).

The concept of the Web as platform is much more than blogging however, for it refers to the idea that gradually we will shift from desktop-based activities to Internet-based ones. Some writers call this "cloud computing" – that software will increasingly move to the Internet where it can take advantage of being massively distributed and collaborational. As such even our operating systems (Windows, MacOS, Linux, etc.) may reside on the Web. For example, Google Apps provide spreadsheet, slide presentation, and word processing tools that enable people to simultaneously work on the same document. Alan MacEachren and his colleagues have also developed sophisticated "geocollaboration" (MacEachren and Brewer 2004; MacEachren et al. 2005, 2006). Notice that this aspect of the Web as platform draws on crowd-sourcing to harness the power of the group.

Size Does Matter!

Any account of the geography of cyberspace (i.e., the Internet and the Web) deserves to begin with a few awe-struck comments about its sheer size. How much human knowledge is there? As we saw earlier (Chapter 5), estimates range from 5 to 281 exabytes. But data (never mind information or knowledge) has a tendency to occur in clumps. As William Gibson (the author of the great cyberpunk novel *Neuromancer*) stated "the future is already here – it's just not equally distributed." This is a clever way of referring to what is generally called the *digital divide*: the difference between the haves and the have-nots in the information economy.

But will universal answering services pose other problems? Will the world be turned into factoids? Will the Internet begin to substitute for lengthier, deeper reflections? Recently, the history department at Middlebury College in New Hampshire decided that it would no longer allow Wikipedia citations in student papers (Read 2007). Others worry about the ability of students to critically evaluate Internet-sourced information, fearing that information obtained over the Internet is inferior or not authoritative (because not peer reviewed). These questions remain unresolved.

Cybergeographies: The Work of Martin Dodge

One of the enduring questions about the Internet is some variant of "What does it look like?" ("Where is the Internet?", "Who is connected to whom?" etc.). At first glance, a map would provide ready answers to these questions. Yet despite the thousands of maps and other visualizations of the Internet, there probably is no satisfactory way to answer these questions, for the following reasons:

1. To conceive of the Internet as a distinct and separate entity ("cyberspace" or the "virtual") rather than a set of heterogeneous processes and capabilities that are inextricably intertwined with our daily lives is probably a mistake.
2. However it is conceived, the Internet is changing not only from day to day, but from second to second.
3. The Internet anyway may not be the best way to conceptualize the total impact of the virtual.

But what does this landscape consist of? We use terms like "cyberspace," "cyber-geography and cybercartography," "the Internet," "the Web," and "the virtual" as if they were all the same thing, but obviously they are not. Some are technical terms for quite specific things – the Web is the network of hyperlinked documents accessible via the Internet, whereas cyberspace is much more nebulous. The term "virtual space" for example was used as long ago as 1953 to describe the spaces created by paintings and artwork (Cosgrove 2005: fn. 11).

None of this has stopped people from attempting to picture and map the Internet. Many of these visualizations have been tracked by Martin Dodge as part of his cybergeography project, particularly his *Atlas of Cyberspace*, an online project he maintained between 1997 and 2004 (Dodge and Kitchin 2001). Dodge is a leading researcher on the geography of the Internet, and with his American colleague Matthew Zook has done important work on the various geographies of the Internet.

Maps in cyberspace

Dodge and Zook distinguish between three kinds of cartographies related to cyber-space: maps *in* cyberspace, maps *of* cyberspace, and maps *for* cyberspace (Zook and Dodge 2009). The first category, maps in cyberspace, is a way of describing how traditional cartography is now available through the Internet – maps are available online and with far greater interactivity than previously. This category essentially sees the Internet and the Web as a publishing and distribution medium, albeit one that allows new kinds of maps to be made. This category is therefore by far the largest of their three categories; covering everything from Google Maps, map mashups, Yahoo, MapQuest, Microsoft Virtual Earth, map libraries whose collections have been scanned, to online and community-based participatory GIS. Anybody or any institu-tion (and there are a lot of them) who has put a map online is part of this category.

Finding these maps has become less of a problem with the development of image search tools. It is an instructive exercise to search for these maps. Say we are inter-ested in cartograms. A text search will give us the description and definition of this map type, but an image search will give us actual cartograms. (In a test I did just now on Google, the work of the geographer Danny Dorling came up in the first page of results.) This makes it very much easier to see if what we are getting is useful or not. It hardly needs to be said therefore that the *size* of the Internet has raised the question of how well search engines work. Just consider this problem for example: which result should come up first in the listing? The most popular – which can be manipulated and may not be the most relevant. The most recent? The most nearby? Ordered by amount of payment to Google? If you are a business which of these schemes would you prefer – and would a user prefer the same scheme? As Zook and Dodge point out, a large business can appear above a nearer but smaller business. In other words, mapping is coming under the control of corporate interests: "control over these maps and the algorithms used to generate them, is vested in private companies without accountability to the public who uses them" (Zook and Dodge 2009: n.p.). It also tends to reinforce an adage as old as the Bible: "to them that hath shall be given, and to them that hath not shall be taken even that which they hath" (Matthew 13:12, also known as the Matthew Principle or effect [Merton 1968]).

Here's a typical experience. Using Google My Maps, I zoom in to Philadelphia, USA, where I am currently staying. As it happens, I am living in a house just around the corner from the Eastern State Penitentiary mentioned by Michel Foucault in his book *Discipline and Punish* which I discussed in Chapter 9. I pop outside and take a picture of it, which I upload into the online photograph hosting site Flickr (owned by Yahoo).

Now using My Maps it is a matter of putting in a placemark over the penitentiary and editing the placemark to include a link to my Flickr photos (and choosing a nice photo to go in the placemark as well). I save the map and make it permanent so that I can share it. In fact, here it is. Type the following URL into your web browser (I've shortened the original google.com URL using a service known as tinyurl): http://tinyurl.com/2as5fu. Now my handmade map has been added to cyberspace! Enjoy.

Maps of cyberspace

Maps *of* cyberspace more directly try to answer the question of what cyberspace actually looks like. Here again there is a startling range of visualizations, from the simple A-----B of the first two computers ever networked (in September 1969) on the ARPANET to the more complex topologies of later years (e.g., MILNET map). Figure 10.1 shows the early structure of USENET, the network that distributed newsgroups (discussion groups, otherwise known as Netnews).

Most of these maps are topological: they show the connections between various computers or computer networks. Sometimes these are mapped over geographical space and can show which areas of the world are well or poorly connected. This point helps us understand why cyberspace (or even the Internet and the Web) are not separate, abstract, and completely virtual systems somehow existing separately from the brute materiality of everyday life. Indeed, it is one of the striking points

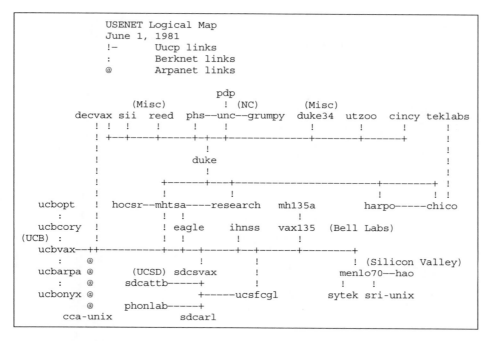

```
              USENET Logical Map
              June 1, 1981
              !-       Uucp links
              :        Berknet links
              @        Arpanet links

                              pdp
              (Misc)           ! (NC)       (Misc)
        decvax sii   reed   phs--unc--grumpy  duke34  utzoo  cincy teklabs
          ! !  !      !       !     !                  !       !      !      !
          ! +--+----+-----+--+-+--+--------------+--------+------+          !
          !                                                                 !
          !                     duke                                        !
          !                      !                                          !
          !            +-------+---+----------------------+--------+ !
          !            !           !                               !        ! !
     ucbopt  !   hocsr--mhtsa----research     mh135a        harpo-----chico
        :    !         !    !               !
    ucbcory  !         ! eagle    ihnss    vax135  (Bell Labs)
   (UCB) :   !         !    !               !
     ucbvax--++----------+--+--+-----+--+------+--------+
        :    @                !         !              ! (Silicon Valley)
    ucbarpa @      (UCSD) sdcsvax       !           menlo70--hao
        :    @      sdcattb-----+       !             !    !
    ucbonyx @             +-----ucsfcgl         sytek sri-unix
             @      phonlab-----+
          cca-unix          sdcarl
```

Figure 10.1 USENET Logical Map, June 1, 1981.

of cyberspace that it is so material and that this materiality has a distinct geography (Crampton 2004).

The use of the word "map" is often metaphorical – it is a topological diagram of a network. Many maps of cyberspace do employ the symbolization of standard maps. A comical rendering is offered by the "Map of Online Communities and Related Points of Interest" (Figure 10.2). Here the size of the feature on the map conveys the size of membership (as of Spring 2007): MySpace, YouTube and the "Blogipelago" are large and dominant, with the Wikipedia project spinning its network (but no Twitter yet) near the Noob Sea (Noob is part of the online slang for newbie or newcomer).

As one person's subjective view of the Internet it is both satirical and informative at the same time, not unlike some of the maps discussed in Chapter 12.

Zook and Dodge point out that few of these maps of cyberspace are made by cartographers or geographers (although Zook created some good ones in his book [Zook 2005]). Most are made by systems analysts because they have access to the data or it is part of their job to predict traffic. This point connects with the claim made in Chapter 1 that a characteristic of modern critical cartography is that it is falling out of the hands of trained cartographers and opening up to other people. This does not mean that everyone is equally able to make maps. If the data remain confidential or held by private companies it is not always so much an infinite widening of mapmaking, but a radical shift.

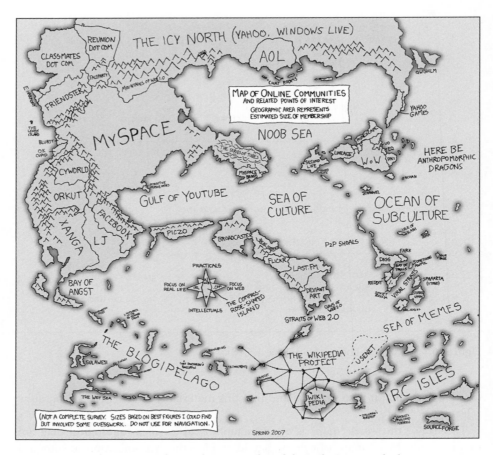

Figure 10.2   A humorous take on the geography of the web. Source: xkcd.com

Maps for cyberspace

Zook and Dodge's final category is maps for cyberspace, which they describe as aids for getting around in cyberspace: "maps become interfaces of exploration 'inside the wires', rather than representations of how the 'wires' themselves are arranged and produced" (Zook and Dodge 2009). By virtue of the fact that these maps provide a visual depiction of abstract relationships, it enables otherwise hidden relationships to be discovered. For example, a blog can contain thousands and even millions of words, but a so-called "tag cloud" can show the most common themes of the blog. Often these might be surprising even to the author of the blog. A tag cloud works by proportionally sizing the word to the number of times it is mentioned. In this example a tag cloud has been generated for the photo hosting website Flickr.com. What are people worldwide taking pictures of? Probably you can predict "friends," "party," and "wedding," but what about "Canada," "rock," and "Japan"? Here we have the mundane, everyday geographies revealed!

All time most popular tags

africa amsterdam animals april architecture art asia australia baby barcelona beach berlin birthday black blackandwhite blue boston bw california cameraphone camping canada canon car cat cats chicago china christmas church city clouds color concert d50 day dc de dog england europe family festival film florida flower flowers food france friends fun garden geotagged germany girl graffiti green halloween hawaii hiking holiday home honeymoon hongkong house india ireland island italy japan july june kids la lake landscape light live london macro march may me mexico mountain mountains museum music nature new newyork newyorkcity newzealand night nikon nyc ocean paris park party people portrait red river roadtrip rock rome san sanfrancisco scotland sea seattle show sky snow spain spring street summer sun sunset sydney taiwan texas thailand tokyo toronto tour travel tree trees trip uk urban usa vacation vancouver washington water wedding white winter yellow york zoo

Figure 10.3 All time most popular tags as recorded at the photographic site flickr.com. Reproduced with permission of Yahoo! Inc. © 2009 Yahoo! Inc. Flickr and the Flickr logo are registered trademarks of Yahoo! Inc.

A more analytical capability is offered by amazon.com and their new capability to map all the places mentioned in a book. Strictly speaking this is a not a map for cyberspace, but it's a map only available because of cyberspace (specifically, Amazon). For instance, a map of all places mentioned in the *Dictionary of Human Geography* (2000 edition) has an almost exclusively Anglo-American geographical orientation.

Why does the book have this specific pattern? No doubt there are good reasons for many of the places mentioned (or not mentioned: it's hard to believe the book almost entirely ignores Africa for instance). Perhaps the technology is imperfect. Perhaps the people who agreed to contribute predominantly came from Europe or North America. Perhaps today human geography is practiced mostly in the West. Perhaps the editors were unaware of many contributors from outside these areas. The point is that the map raises these questions about the inequalities of knowledge distribution. In the last section of the chapter, I will look at this question in more detail, and discuss the role of "net neutrality."

The Digital Divide

Consider a map of Internet access created by digital artist Chris Harrison using data from a "crowdsource" effort known as the DIMES project (Figure 10.4). Harrison's

ChrisHarrison.net

Figure 10.4 Global internet connectivity, by Chris Harrison (left). Detail of Europe (right). Source: Chris Harrison, Carnegie Mellon University. Used with Permission.

Figure 10.4   (*Continued*)

map shows the degree of connectivity around the world – the darker the symbol, the more connectivity. It quickly becomes apparent that the distribution of access ("on-ramps") to the Internet is very concentrated.

Although the map does not show any country outlines, it is fairly easy to discern that North America (particularly the USA) and Europe dominate. Even within these areas however, access is patchy (e.g., the American Midwest), probably because fewer people live there. South America and Africa barely appear, and even then mostly around the coastlines. Japan stands out strongly, as do the southeastern coastal areas of Australia.

As of 2005, there are just 18 countries in the world where more than half the country's population can access the Internet (United Nations Development Program 2007). High-income OECD countries average 52.5 percent access; developing countries average 8.6 percent as a whole, with the least developed countries averaging just 1.6 percent access to the Internet. These concentrations or divides occur at multiple scales: globally, regionally, and locally. Globally, Internet access rates are concentrated in a few countries you might expect (the USA, UK, and Western Europe), as well as some that you might not (the best connected country in the world? Iceland, with 87 percent online). Scandinavian countries all have better access rates than America or Britain. Conversely, there are many countries with extremely poor rates of access, such as almost the entire continent of Africa, but there are also surprisingly low access rates in highly technologically advanced countries such as Japan and France (8th and 10th in the world respectively).

When we consider the world-wide distribution of other resources, such as clean potable water, the geographical patterns are remarkably similar. Where people lack even basic necessities such as water, they lack Internet access. But Internet access provides a particularly sharp reminder of the structural obstacles to human development. Whereas many of the places suffering from lack of clean, inexpensive water have to some extent traditionally suffered from that problem, the newness of the Internet – the fact that just a decade ago hardly any country had access to it – shows how inequalities are continually produced. In 1990 for example, the highest Internet access rate was just 8 percent (in the USA), only eight countries had Internet access rates above 1 percent, and only 14 had any measurable access *at all* (United Nations Development Program 2006: Table 13). Beginning as it were then from this common zero line, 15 years later the map shows stunning inequalities in access, with runaway connectivity in the West, while sub-Saharan Africa's best-connected nation (South Africa) clocks in at a mere 10.9 percent in 2005 – in other words, slightly higher than that of the USA in 1990, nearly two decades ago.

This concept of unequal access has been dubbed the "digital divide" which can be defined as "unequal access to the information economy." Note that this is not primarily a question of technology, as is commonly assumed, but one of knowledge: knowledge of how to use the technology, of education in the information economy, *plus* the sheer access to the technology. The result of this knowledge-based approach (sometimes called "access to knowledge" or A2K) is that it first becomes very evident that the digital divide is not something that can be overcome by improving access to technology itself (although that will help). The frequent announcement of the "end" of the digital divide following the development of some new technology is unwarranted. Not only is there no associated training to go with this new technology, but also little indication of how it will be helpful for populations living on a few dollars a day.

Over the past five years for example, the One Laptop Per Child (OLPC) initiative has developed a robust and inexpensive laptop that may sell for as little as $100 (at the time of writing, the BBC reports it costs $176). The first machines from this effort were scheduled for delivery in late 2007. The project is the initiative of Nicholas Negroponte of the Massachusetts Institute of Technology, and it represents a marvel in inexpensive computing. The first models of the laptop (called the XO) sport a color 7.5 inch screen, a 433 MHz processor, 1 gigabyte of flash memory, and wifi capabilities. The problem of the lack of electricity is addressed through the options of a basic rechargeable battery, a hand crank, and a pull-string recharger, which gives ten minutes of use for every minute of pulling. Whatever you think about it, the engineering behind this computer is impressive.

But the machine is not without its critics. Bill Gates criticized its "tiny screen." Intel's CEO accused OLPC of making a "gadget," when what people really want is a fully functioning modern PC, and implied that it was discriminatory to offer this computer to developing countries, while developed countries have so much better machines (Intel has since backed away from these criticisms and is now working with OLPC). The laptops are meant to be sold to governments and then distributed to children, but some NGOs argue that government money would be better spent on clean water and schools – a $2,000 library can serve 400 children at $5 each, argued John Wood of Room to Read, a nonprofit organization that promotes literacy in developing countries:

"These kids in rural Cambodia can't even read yet," Wood says. "*What are they going to do with a computer?*" A small rural library serving 400 kids costs $2,000 to set up. Five dollars a child. Computers are far more expensive. There are places and times when computer labs can be helpful, Wood says, and Room to Read will fund about 30 this year. But "we'll do 900 libraries. We'll do about 85 new school construction projects." (Thompson 2006: emphasis added)

If his figures are right then 900 libraries will potentially service 360,000 children at a cost of $1.8 million, or $1/20$ the cost of the OLPC.

Secondly however, the divide is really rather a series of gaps, not one big gap. Technological innovations occur in waves, and while it may be the case that eventually the majority of people will adopt a certain innovation, by the time they have done so a new wave of innovations is coming along which they do not have. Consider the modem. The earliest modems (in the 1960s) operated at 300 bits per second. By 1980 some modems were capable of 14.4 kilobits per second (kps). From there speeds increased to 56 kps and now cable modems and DSL ("broadband") operate at speeds of 3–8 megabits per second.

This might not constitute a problem if information was designed or even compatible with the lower end of the market, but the fact is that anybody who is connecting via dial-up today will be very frustrated by the large file sizes and operating requirements such as RAM that are needed simply to surf the Web. The same is true for running mapping or GIS software. ESRI's stated requirements for running ArcGIS 9.3, for example, call for a CPU of at least 1.6 GHz, 1 GB of RAM, 2.4 GB of disk space, as well as high-end add-ons from Microsoft such as .NET Framework 2.0 which carry their own burdensome requirements.

Thirdly, the digital divide is not just a technological issue because it is more importantly one of justice and equality. Jack Balkin, a professor of law at Yale with interests in information access, has outlined the stakes as follows:

1. Human knowledge – education, know-how, and the creation of human capital through learning new skills.
2. Information – like news, medical information, data, and weather reports.
3. Knowledge-embedded goods (KEGs) – goods where the inputs to production involve significant amounts of scientific and technical knowledge. Examples include drugs, electronic hardware, and computer software.
4. Tools for the production of KEGs – examples include scientific and research tools, materials and compounds for experimentation, computer programs and computer hardware. (Balkin 2006)

These knowledges are much harder to map than Internet access. More than one measure of knowledge development likely would be needed. Many of Balkin's forms of knowledge would be delivered not only via the Internet, but also by radio, newspaper, parents, the community, or classroom.

Simple cyberspace maps of the topological connections then, are not the whole, or even the most important part of the story. A recent attempt to provide a better insight into the shape of the Internet was attempted by some Israeli researchers (Carmi et al. 2007). The DIMES project attempts to give a better picture of how knowledge may be distributed. One of their findings is that a significant proportion of the network is composed of fairly isolated nodes that can only reach the rest of the network through the central core. In urban terms, this is analogous to "edge cities"; those exurban areas outside the downtowns that are well connected and serviced with facilities for shopping and working. The isolated outer areas are like widely scattered small towns that can only connect with each other through an airline hub.

Finally, we should consider another sort of uneven access which goes by the name net neutrality. If the Internet is conceived as a hierarchy, as Carmi et al. argue, then this hierarchy is not solely produced by the differential physical connections between core and periphery. Rather, it can also be produced deliberately. Telecom companies have argued that such a hierarchy should be produced through differential pricing for websites. That is, websites should pay more for higher speed access, for having more computers connected, or for having certain types of content. Since the early 2000s advocates of net neutrality have campaigned against the telecoms, arguing that loss of net neutrality will result in highly differential access. Some of the very figures who originally developed the Internet are among these advocates. Vint Cerf, for example, has been quoted as saying that "the Internet was designed with no gatekeepers," and Tim Berners-Lee, who invented the Web itself, is also opposed to it, along with most of the major Internet companies such as Google, Amazon, Yahoo, and Microsoft. As I mentioned earlier in the book, blog activists have been a critical part of the opposition to telecoms, and their efforts (and a change in Congressional power in 2006) have so far stymied attempts to introduce anti-neutrality legislation.

We have seen in this chapter that the concept of "mapping cyberspace" is a complex one. Yet at the same time these very attempts to map cyberspace reflect our yearning to come to terms with it, to struggle with it and over it. From its origins in the 1960s as a research/military application, the Internet has become increasingly commoditized and at the same time a key development in globalization. If most of the world's countries still have poor access to the Internet, the lives of their citizens have nevertheless been touched by it. Whether it be because companies in developed countries are outsourcing business more flexibly (e.g., help centers in India) or because of ever-finer surveillance systems that can track and map human and environmental changes, there is less and less opportunity to escape the information economy. How these structures and processes are revealed, and how the relations of power and knowledge are produced, remains a question that critical cartographers and GIS users will need to continually address (see discussion in Chapter 7).

Note

1   Jimmy Wales, the founder of Wikipedia, objects to the term crowdsource, arguing that it does not apply to Wikipedia because it demeans the work of the contributors. That is, it tricks people into working for free and exploits their labor. However, there is nothing inherent in crowdsourcing that means it could not reward labor, although whether this would occur at "fair" market rates is open to question.

Chapter 11

The Cartographic Construction of Race and Identity

Introduction

There is a storied relationship between mapping and the construction of race and racialized territories. A racialized territory is a space that a particular race is thought to occupy. In this chapter we'll look at the ways in which mapping and race have intersected. We'll also try and clarify the changing ways in which the notion of race itself has been understood – and conclude by noting some of the ways that older ideas of race as biological are unfortunately reemerging.

Efforts to spatially characterize human populations extend back into antiquity. Strangeness could be located – it was those who were away from you. In his *Histories* (written around 430 BC) Herodotus did not hesitate to describe the various peoples in far-flung parts of the "oikumene" (inhabited world), however strange they were.

> These men it would seem are wizards; for it is said of them by the Scythians and by the Hellenes who are settled in the Scythian land that once in every year each of the Neuroi becomes a wolf for a few days and then returns again to his original form. (Hdt. 4.105)

Herodotus wasn't gullible, he immediately adds that this story is a bit of a stretch. But it illustrates the problem: he had to rely on secondhand knowledge and sailor's stories. His geographical locations were general and there was no attempt to make precise boundaries. Though peoples were sometimes described as occupying particular locales, these were not political borders.

Prior to the age of exploration most Europeans had little knowledge or experience of "other peoples," but if they did think of them it was in terms of away (even if "away" meant the next village). In medieval Europe, for example, a whole series of maps were produced that literally marginalized odd and different peoples, if not as races then as fabulous or biologically different in some way. The Hereford mappa mundi (map of the world) of c. 1290 AD shows a bountiful array of humans, half-humans, dog-headed men (cynocephali), and monsters, especially around its circumference in the East and North.

The map is in the distinctive circular shape bisected on the interior with the Mediterranean, the Nile and Don rivers in a T shape, known as T-and-O maps. It is oriented with East at the top (the most holy direction and the location of the Garden of Eden). Western Europe is drawn more or less correctly, centered around Jerusalem. The map-makers did not think that Jerusalem was literally the center of the world, but wished to emphasize its religious importance by putting it in the middle. Around the margins we can see dog-headed people, men with ears as big as their bodies within which they could wrap themselves for warmth, an illustration of a man riding a crocodile in the Nile, the golden fleece from the story of Jason and the Argonauts, cannibals, and men without heads but with eyes in their chests. Many of these ideas were derived from classical authors such as Pliny the Elder whose 36-volume *Natural History* (1st century AD) described the "blemmyae" (the headless people with eyes in their chests), "sciopods" or shadow foots (their name derives from the fact that they had one huge foot at the bottom of their fused legs, under which they could shelter if they lay on their backs; this creature was resurrected by C. S. Lewis in his *Chronicles of Narnia* stories), and the "cynocephali" (dog-head) (Winlow 2009). These "Plinian races" were well known to later writers and were often repeated in their own accounts, including many medieval maps.

Pliny says that in India one may find:

A tribe of men called the Monocoli who have only one leg, and who move in jumps with surprising speed; the same are called the Sciapods, because in the hotter weather they lie on their backs on the ground and protect themselves with the shadow of their feet. (Pliny the Elder 1938–63: Vol. 7:23)

Friedman (1981) suggests that these strange races were possibly derived from medieval observations of people in yoga positions! (See Figure 11.1).

Medieval mappae mundi divided the world into three partitions, according to the Biblical tradition in Genesis (Chapter 10) that the three sons of Noah (Shem, Japheth, and Ham) peopled the earth (Wallis and Robinson 1987). Thus the world was populated by three major groupings of peoples (Asians, Europeans, and Africans) which could later be interpreted as distinct races. For hundreds of years maps showed this tripartite division of the separate geographical origins of human races (called "polygenism") which was sometimes used to show the supposed superiority of white races, especially if it could be coupled with preadamism. Noah's curse of his son Ham in Genesis (9:25), along with the fact that he is allocated to Africa (and perhaps that his name was often confused in medieval writing with that of Cain, also cursed see Friedman 1981), provided grounds not only for prejudice against Africans but justified their slavery.

The increasing sense in the West of an "us" and "them" can be traced back to the encounters produced following the great age of discovery after the fifteenth century. According to the anthropologist Jonathan Marks this sense of the "other" gradually expanded from a quite local one, perhaps of the village in the next valley, to

Figure 11.1   Illustration of monstrous races in India from Sebastian Munster *Cosmographiae Universalis* 1544 (page 1080).

continent-wide designations during the nineteenth century, that is, to the idea of geographical races (Marks 1995; 2006). As Winlow has discussed, the establishment of evolutionary theories in the nineteenth century served to redouble efforts on mapping human racial types (Winlow 2009), as part of a whole concern with human characteristics, population density, migration, longevity, and especially language and religion. These latter two could and did often act as surrogates for "race."

By the mid-nineteenth century multiple forms of mapping were in use to understand human populations. The subject matter of these maps was initially population density, but more refined understandings of human population groups were soon developed. Maps were made of mortality, education, crime, longevity, language, religion, birth and death rates, age of first marriage, and so on. These subjects were of concern as "moral statistics," or how best the modern state should be governed (Hacking 1982) as we saw in Chapter 6.

Changing Conceptions of Race

None of the early writers noted above thought of the monstrous and strange peoples or creatures in terms of race as we understand it today. Indeed the very idea of race has fluctuated and changed and it is important to understand how it has done so.

Race (and racism) today is largely a product of the last three centuries. Marks observes that during the twentieth century alone there were at least three ways of thinking about race (Marks 2006), which he summarizes as follows:

1. Race as essence. In the early twentieth century race was something that was in you. Heavily influenced by Mendelian genetics. "One drop" rule. Major problem: "passing" or appearing to be something you are not.
2. Race as geographical populations. From 1930s onwards. You are part of a race. Major problem: physical features grade continuously across space, making it hard to unambiguously identify races.
3. Modern genetics. Genetic distinctiveness (however tiny) of groups of people. Problem: extremely weak explanatory power, often accounting for less than 5 percent of variation.

As can be seen from this breakdown, if you wish to maintain the reality of race there is more than one way to go about it, but these ways are not scientific or inferred from biological evidence. Rather, they are cultural constructs.

This can be further demonstrated by looking at the way race has been included in national censuses. Since the American census was first collected in 1790 for example (the UK census began shortly afterwards, in 1801), the number and definition of races has changed almost every time (Table 11.1).

Table 11.1   Changing categories of race in the American census at selected dates.

Date of census	Racial and ethnic categories used
1790	Free white males and females, All other free persons, Slaves
1820	Free white males and females, Free colored persons, All others except Indians not taxed, Slaves
1870	White, Black, Mulatto, Chinese, Indian
1890	White, Black, Mulatto, Quadroon, Octoroon, Chinese, Japanese, Indian
1930	White, Negro, Mexican, Indian, Chinese, Japanese, Filipino, Hindu, Korean, Other
1960	White, Negro, American-Indian, Japanese, Chinese, Filipino, Hawaiian, Part-Hawaiian, Aleut, Eskimo
1980	White, Negro or Black, Japanese, Chinese, Filipino, Korean, Vietnamese, Indian (Amer.), Asian, Indian, Hawaiian, Guamanian, Samoan, Eskimo, Aleut, Other
2000[1]	Spanish/Hispanic/Latino, White, Black, African Am., or Negro, American Indian or Alaska Native, Asian Indian, Chinese, Filipino, Japanese, Korean, Vietnamese, Other Asian, Native Hawaiian, Guamanian or Chamorro, Samoan Other Pacific Islander, Other

Source: Adapted from Tables 1 and 2 in Nobles (2000), US Census Bureau.
[1] The 2000 census was the first to allow respondents to select more than one racial category and to combine this with the Spanish/Hispanic/Latino category.

It is apparent that racial identity has become more fractured in America over time, with just four categories in the first census and 15 in the last one. Although Americans were offered the chance to identify as multiracial for the first time in 2000, and even to combine this with the ethnic group "Spanish/Hispanic/Latino" if they so wished, less than 2 percent of Americans selected this option. While due to the size of the United States this amounts to millions of people, the fact remains that most Americans today still see themselves belonging to a single racial category. For the modern student of race then it is understood that there is no biological basis for racial classifications and that physical variation occurs continuously over space with no hard and fast boundaries. This is quite opposite from the mutually exclusive categories and solid cartographic lines of most race-based mapping. Such an understanding however by no means implies that people have not searched for them.

Race-Based Mapping

As we saw in Chapter 6 the idea that humans can be assigned to a small number of distinct populations got a major boost from the work of Carl Linnaeus, whose mid-eighteenth-century *Systema Naturae* (10th edn.) was highly influential. Linnaeus's four racial categories – blue-eyed white Europeans, kinky-haired black Africans, greedy yellow Asians, and stubborn but free red native Americans (the descriptions again are his own) – were natural categories and if there were exceptions it was only because they differed from an ideal Platonic type. If there were "abnormals" this was because they (perversely, delinquently) deviated from the normal (Foucault 2003a). Linnaeus appended a "monstrous" category to his original four to account for these abnormals (*Homo sapiens monstrosus*), such as the Patagonian giants, the dwarfish Alpines, the cone-heads of China, and so on. These were clearly marginalized figures who existed well beyond the bounds of the familiar world.

With the invention of thematic mapping in the late eighteenth century it became possible for the first time to coherently map these races and describe their geographies. Although Herodotus, Pliny, Sebastian Munster, and others could roughly locate these races this was only in a vague geographical sense. If until the eighteenth century most maps were general reference maps or topographic depictions, thematic depictions of one specific topic (such as population density) began to appear in the late 1700s. It is not surprising therefore that at that time we see race-based mapping flourishing, especially in France where many forms of thematic mapping were developed (Konvitz 1987).

If the focus of the state was increasingly on populations then maps are particularly apt descriptors of groups. While one certainly can map an individual (as was sometimes done in tracing the route of the great explorers around the globe, such as Magellan), this is a practice of the surveillant state. Even today it is more typical to group or agglomerate individuals together for a synoptic view than to map individuals. Thus maps and the politics of populations went together. But what kind of maps and what kinds of populations did they show?

There have been two major ways to understand populations and their territorial distribution. In one kind, typified by choropleth maps, the density or degree of a feature is mapped in pre-given political units. These units can be any arbitrarily defined political entity, including census blocks, census tracts, counties, states, or countries. A second kind of map, known as isarythmic mapping by geographers and clinal maps by anthropologists, shows change varying continuously over space. An isarythmic map is analogous to a map of the earth's surface (a contour map is an example). These changes may be fairly smooth or, as with the earth's surface, we may see fairly steep slopes where attributes change quickly over space. What is critical about these two approaches is that they produce different ways of understanding human variation. Where choropleths produce a sense of populations as contained within boundaries, isarythmic mapping emphasizes continual variation and gradual change without clear differentiations. During the nineteenth century discrete racial types were linked to specific territories, often using choropleth and similar maps. It is no surprise therefore that the history of race-based mapping has continually favored bounded groupings rather than the clinal variation we accept today.

Nevertheless, other types of maps than the choropleth were certainly available. Early clinal maps are known from at least 1701, when Edmond Halley produced a thematic map using isolines to show magnetic declination across the globe, and a century or so later Alexander von Humboldt presented a more refined technique of "isotherms" (Robinson and Wallis 1967). We should also note that a concept known as the "isocline," or line of equal slope, was developed at this time (early nineteenth century), and is still today used in population dynamics.

Maps were also made where different peoples came together, or possibly were encroaching. For example, Ami Boué's 1847 map showed the distribution of "volk" across Europe with a focus on the Ottoman Empire in Europe. Boué, an Austrian geologist by training, went to Edinburgh in the aftermath of the Scottish Enlightenment and studied under Robert Jameson.

Similarly, Gustav Kombst's mid-century ethnographic map in A. K. Johnston's *Physical Atlas* showed various racial groupings allotted into distinct geographical territories (see Figure 11.2).

These maps were not just made for the sake of it. They entered an increasingly racialized discourse that was concerned with distributions of populations (often, as in Kombst, to compare the state of "pure" races with encroachments from other, lesser races). As the century wore on, distinct racial categories became more, not less pronounced. By the start of the twentieth century, less than a hundred years ago, the doctrines of eugenics could be not only supported but were actually mainstream. The center of this was the Eugenics Record Office (ERO) in Cold Springs, New York, which was supported by leading philanthropists and the Carnegie Institution of Washington.

Meanwhile the USA was off to an uncharacteristically slow start. It produced no statistical atlases for most of the nineteenth century, but after the Civil War

Figure 11.2   Above: Gustav Kombst racial map of Europe, 2nd edition 1856. Right: detail of the UK. Used by permission of David Rumsey Map Collection (www.davidrumsey.com).

the need for an accounting of the various populations became too difficult to ignore. The Census, which up to that time had been dominated by patronage and nepotism, got a professional upgrade with the new Superintendent, Francis Amasa Walker (we previously him met in Chapter 6). Walker's new professionally based maps reveal distribution after distribution of ethnic, racial, and national groupings across the USA. The biggest one occupied a two-page spread and was therefore some 28" across – more than 2 feet showed the "constitutional" population (see Figure 6.2). But this was followed on by maps of the colored population (see Figure 11.3), a map of those with foreign origins, those who were native born but whose *parents* were foreign-born, and then a whole series of smaller maps showing distribution by nationality: Irish, German, English and Welsh, and Scandinavian. It was a tremendous success and Walker had enough copies of the *Atlas* printed so that it could distributed to schools for the edification of the nation's young citizens.

But of course the point was not just education. Writing in the *Atlantic Monthly* in 1896 Walker argued that immigration restrictions should apply not only to

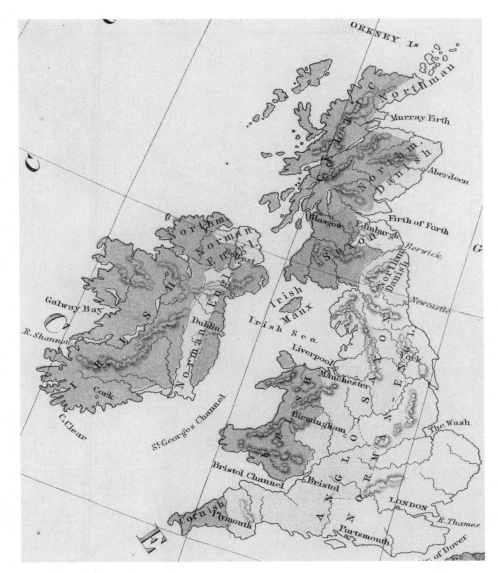

Figure 11.2 (*Continued*)

hundreds or even thousands of people but to hundreds of thousands, and not because they were deaf, dumb, blind, or criminal, but simply because they would subject America to "degradation through the tumultuous access of vast throngs of ignorant and brutalized peasantry from the countries of eastern and southern Europe" (Francis Amasa Walker 1896: 823; see also Sluga 2005). In other words the quality of the American population would be reduced by what was at the time the commonly identified threat of southeastern Europeans (Italians, Slavs, Greeks, Hungarians, and so on).

Figure 11.3 Map of the colored population of the USA. Source: Francis Amasa Walker (1874).

Where Linnaeus had classification as his goal (isolation of common elements), Buffon had *diversity* as his (explaining variation). Unfortunately, says the anthropologist Jonathan Marks, between 1758 (Linnaeus's 10th edition) and the 1960s (when it was finally overthrown), physical anthropology followed Linnaeus and not Buffon in searching for races and their nature:

It is one of the blindest alleys in the history of modern science. The question ignores the cultural aspect of how the human species is carved up; it ignores the geographically gradual nature of biological diversity within the human species [i.e., clines] and it has a strong anti-historical component in its assumption that there was once a time when huge numbers of people, distributed over broad masses of land, were biologically fairly homogenous within their group and different from the (relatively few) other groups. (Marks 1995: 52)

In other words it is clines and not choropleths. The choropleth map has contributed to this blind alley, whereas understanding the range of biodiversity in human populations allows us to see how they vary gradually with environment, migration, and genetic drift.

"A hatching oven for war": the cartographic calculation of race

As an illustration of this point we can briefly consider an important work carried out just prior to the First World War by a member of the American Geographical Society (AGS), Leon Dominian (1917). Dominian's book was an attempt to show the relation between language and nationality with a view to settling political boundaries. An "ill-adjusted boundary is a hatching oven for war. A scientific boundary . . . prepares the way for permanent goodwill between peoples," he begins (1917: vii). Dominian pays particular attention here to the "Eastern Question" (the territorial problems posed by the decline of the Ottoman Empire), and how geographical knowledge could provide an acceptable settlement. This book was part of an increasing shift from a geopolitics dependent solely or mainly on "natural borders" (ridge lines, rivers, watersheds) that could provide defensible boundaries, to one that increasingly incorporated "population borders" whether using race, language, religion, economic trade, and so on.

Dominian used both, arguing that borders began in nature but were elaborated by humans, and that natural borders then fade out as "the result of man's progress . . . [by] the removal of natural obstacles; the conquest of distance by speed" (1917: 327). Dominian himself highlighted economic development, and it is the more remarkable then that his book should have an Introduction by Madison Grant, author of perhaps one of the most notorious racist tracts of the early twentieth century (Grant 1932: 1st edn. 1916). In fact Grant was an AGS Councilor for several decades and his book first appeared in its journal the *Geographical Review* (Grant 1916). For Grant race was a meaningful biological (phenotypic) variable: "race taken in its modern scientific meaning [is] the actual physical character of man" (Grant 1917: xv), and "it is entirely distinct from either nationality or language . . . race lies at the base of all manifestation of modern society (Grant 1932: xxi). Race, for Grant, was a substrate written into human biology. It is neither a linguistic nor a political group (Grant even observes this in Dominian's book, warning him off seeing race in his linguistic groups). Even achievements made by non-whites were the result of "mimicry" of whites imposed from without by social pressure of the "slaver's lash":

> Whenever the incentive to imitate the dominant race is removed the Negro or, for that matter, the Indian, reverts shortly to his ancestral grade of culture. In other words, it is the individual and not the race that is affected by religion, education and example. Negroes have demonstrated throughout recorded time that they are a stationary species and that they do not possess the potentiality of progress or initiative from within. (Grant 1932: 77)

Race then is inherent and fixed, biological. Grant insisted that there were three major races in Europe (Nordic, Mediterranean, and Alpine). Outside Europe the major races were "Negroid" and "Mongoloid" (1932: 32). Some countries were more affected by archaics with "Neolithic" traits. In Britain for example, while admirably Nordic in general (blond, blue-eyed, flowing hair), it sometimes yielded evidence of this less-developed trait. Who can fail to observe, says Grant "on the streets of London the contrast between the Piccadilly gentleman of Nordic race and the cockney coster-monger of the old Neolithic type" (1932: 29)? Grant even attempted to portray differences between the sexes in terms of differential evolution: women exhibit "the older, more generalized and primitive traits of the past of the race" (1932: 27).

Grant vociferously denied that this innate substrate could be molded by the environment, which as we have seen is a central tenet of the explanation for genetic changes in a population. There is "a widespread and fatuous belief in the power of the environment, as well as of education and opportunity to alter hereditary, which arises from the dogma of the brotherhood of man" (1932: 16), he wrote. Grant sarcastically makes fun of emerging anthropological findings that even head shape could change among immigrant groups, a finding first discovered by Franz Boas and now generally accepted in anthropology.

As was written in his own book's Preface:

> . . . if I were asked: what is the greatest danger which threatens America today? I would certainly reply: the gradual dying out among our people of those hereditary traits through which the principles of our religious, political and social foundations were laid down and their insidious replacement by traits of less noble character. (Grant 1932: ix)

Dominian's book itself not surprisingly deploys many maps that attempt to delineate the geographical extent of various languages (see Figure 11.4). His task here may seem quite a daunting one given the fact that the predominant language of a region may not be the only language of a region, and that dialects within a language add an additional complication. But Dominian, although perhaps not in a position to point out differences with his employer's Councilor, does not take the same approach as Grant. For Dominian, what is important is not so much the inevitability of the racial substrate, as the effects of the environment, of economics, and of human development; the non-biological. Although he believes heredity is important he takes a much wider approach in understanding human variation, perhaps what we would today call one of "nature-culture" (Goodman et al. 2003). Nevertheless this was still an exercise in finding bounded regions with a view to establishing "scientific" political borders.

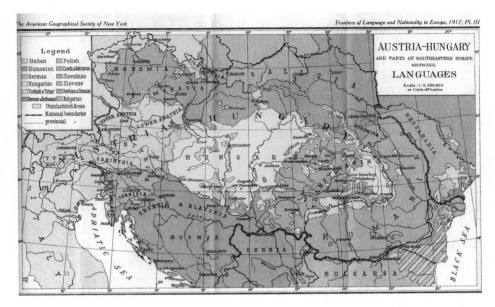

Figure 11.4 Dominian's mapping of language distributions in the Austro-Hungarian region during World War I. Source: Dominian (1917).

The Reinscription of Race?

Scientific eugenic theories were finally put to rest in the United States with the coming of World War II. The ERO was disbanded on the eve of war when the Carnegie Institute could no longer justify its research. Following the war the United Nations Educational, Scientific, and Cultural Organization (UNESCO) issued a number of Declarations on race. These statements rejected the biological basis of race, and clearly broke with the key tenets of Nazi racial theory that had stayed their hand before the war for fear of antagonizing Hitler. The 1950 statement read in part:

> From the biological standpoint, the species *Homo sapiens* is made up of a number of populations, each one of which differs from the others in the frequency of one or more genes. Such genes, responsible for the hereditary differences between men, are always few when compared to the whole genetic constitution of man and to the vast number of genes common to all human beings regardless of the population to which they belong. This means that the likenesses among men are far greater than their differences. (quoted in Graves 2001: 149)

In other words race is not a valid biological category for scientific research. This has now been the accepted position among scientists for five decades.

In the last few years however there has been an attempt in some scientific quarters to bring back biological explanations of race. The sociologist Troy Duster describes this as the biological "reinscription" of race (Duster 2005). Much of this biological reinscription comes from the biological and medical communities, including human genetic research.

A notable example of this reinscription occurred in 2005 when the US Food and Drug Administration (FDA) gave approval for the first so-called "race-based medicine," BiDil (Kahn 2007; Sankar and Kahn 2005). BiDil was aimed at treating congestive heart failure among African Americans, and the company that manufactures the drug, NitroMed, observed that "African Americans between the ages of 45 and 64 are two and a half times more likely to die from heart failure than Caucasians in the same age group" (NitroMed Inc. 2005). It was widely hailed as a new breakthrough in medical science's ability to target drugs at specific populations (known as "pharmacogenomics"), and to provide better health care for an underserved population, African Americans (see Figure 11.5).

Unfortunately the reality was more complex. First, the drug was in fact not new, rather it was a combination of two existing drugs into one pill, the generics hydralazine and isosorbide dinitrate, both of which have been around for more than a decade at one sixth the cost of BiDil. Second, there is no firm evidence that BiDil works any

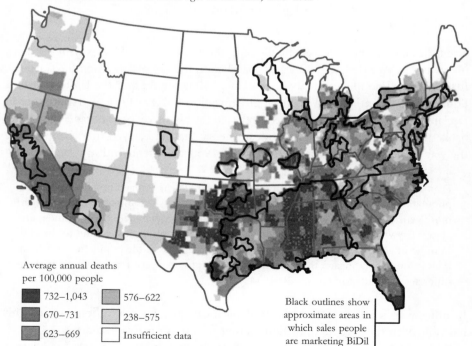

Figure 11.5 BiDil, the first "race-based drug" approved by the FDA. Source: AP. Used with Permission.

differently for African Americans than any other population – the FDA approval was based on a second experiment that only used African Americans and did not compare results to other populations who might benefit from it. Third, the age group 45 to 65 accounts for only 6 percent of heart failure mortality, and for people over the age of 65 there are almost no statistically significant differences between African American and Caucasian mortality rates (Duster 2005). In studies comparing whites and blacks from various European, African, Caribbean, and American populations, it was found that hypertension rates showed a significantly smaller racial disparity for populations from Brazil, Trinidad, and Cuba than in North America. Such a study points strongly to a socio-cultural predictor of health rather than a biological one.

Nonetheless race keeps coming back whether it be in the racial account of intelligence offered in the infamous *Bell Curve* book, health and medicine, or genetics. In 2005 when the *New York Times* published an op-ed by Dr Armand Marie Leroi, an evolutionary biologist at Imperial College, the Social Science Research Council, perhaps recalling the purchase that the *Bell Curve* obtained on the public consciousness, organized a series of 13 responses and placed them online (http://raceandgenomics.ssrc.org/). Leroi argued that longstanding (post-war) understandings of race as a social construction were being outmoded by new results from genetics, which was successfully identifying biological races.

The responses were garnered from an interdisciplinary panel of writers under the heading "Is Race 'Real'?" As the discipline that has made a profession of race studies for over two hundred years, the responses from anthropologists were the most useful (no geographers were invited). Alan Goodman's response dwelt not only on refuting Leroi's argument that race has a biological basis, but also the concomitant assumption that it therefore does not exist as lived reality at all (Goodman 2006). Race and racism need not be biological to be a lived experience for millions of people. Nor does it mean that human variation is not real; obviously people differ biologically – but these differences do not resolve to races. What differences we see are distributed gradually (as noted above, anthropologists, ecologists, geneticists, and others use the word "cline" to denote these gradual geographical variations, and although in geography it is possible to speak of isarithmic changes, cline would be a useful addition to the geographer's lexicon, see (Crampton 2009b).

Jonathan Marks made a related point about the geographical reality of races that is worth quoting in full:

> Such an assertion of qualitative geographical distinctions – race as continent – is not natural, not objective, not value-neutral, not scientific, and not being inferred from the data. It is, rather, the artificial division of a continuum into discrete sections, and the imposition of meaning or significance upon the separation between those sections. (Marks 2006)

Both writers cite the definitive work of Lewontin, who showed in the early 1970s that genetic differences *between* racial groups was far smaller (6.3% of variation) than that observable within a race but among local groups (85.4% of variation)

(Lewontin 1972). In other words if you're looking for biological differences, you can't find them between races.

Some Remaining Questions for Geography and Race

These developments have implications for the study of geography from a racial perspective. Although a number of disciplinary organizations have issued statements on race and its scientific study, including the American Anthropological Association (AAA), the American Association of Physical Anthropologists (AAPA), and the American Sociological Organization (ASA), similar statements have not been issued by geographical associations such as the Association of American Geographers (AAG) or the Institute of British Geographers (IBG). This is despite the fact that geographers are significant consumers of race-based data such as that collected in the census. In this way professional geographers have absented themselves from participating in one of today's most critical social issues.

We need not remain silent. There are a number of questions that would benefit from geographical insight. Should race-based categories continue to be used in geographical studies? In this context it might be noted that some countries such as France have stopped collecting race-based data. Some have argued that if we do stop collecting race-based data it will mean we are no longer able to track discrimination. For example, we would not be able to count the number of African Americans living below the poverty line, or who have access to good health care, and to map out where they live. In France these objections have been answered by recording the more fundamental data of the number and location of people living below the poverty line. They argue that ultimately it is poverty (or health, etc.) that matters, whoever it is that is affected. Thus the country that in the nineteenth century introduced many of the thematic maps used in today's GIS now offers an example of what happens when you fundamentally change the categories of your data. Critics of this change point out that racism is still prevalent in France, although whether this is caused by racial categories or its history of colonialism is not clear.

Geographical studies have tended to focus, and rightly so, on the nature of racism and discrimination. Most geographers would probably stipulate that racism is objectionable, but would they also stipulate that racism is predicated on racial categories and that by employing racial categories they are reproducing and sustaining them? Why have we as geographers not studied *race* as much as we have studied *racism*? Is it because we are happy to use the racial categories handed down to us from the census and other stored databases (e.g., DNA-based crime databases)? But as we have seen, these categories change markedly over time, and in the case of crime databases may be subject to racial and ethnic bias. Why are we satisfied with the categories of stored data sets and what does our usage of them entail for our results? Would we appreciate some input into the formation of those categories?

Why has the profession not issued a public statement on race-based data in, for example, the 2010 Census?

What about the refusal of most Americans to see themselves as having a mixed heritage: was this option not chosen due to its unfamiliarity on the 2000 Census and if so, how can geographers play a role in making that option more easily understood? Was it because of common misunderstandings about race that it is fixed and mutually exclusive, and again if so, how can geographical insights into human biodiversity as a continuous geographical variable help to clarify what race means? For example, after every census communities undergo a redistricting process which adjusts the boundaries of the political maps. Legally, as long as these boundaries are not capricious race can be one of the variables politicians use to redraw the maps – but what notion of race and its geographical distribution is being employed here?

Finally, why is race data collected from every household in the US census but data on poverty and income collected from only 17 percent of households?

None of these questions admits of an easy or even definitive answer. But it is significant that geographers, for all our technological GIS capabilities, have not even begun to ask them.

Chapter 12

The Poetics of Space:
Art, Beauty, and Imagination

Space seems to be either tamer or more inoffensive than time: we're forever
meeting people who have watches, very seldom people who have compasses.
   Such places don't exist, and it's because they don't exist that space becomes a
question, ceases to be self-evident, ceases to be incorporated, ceases to be appro-
priated. Space is a doubt.   Georges Perec (1974/1997).

So far in this book we have considered mapping without much discussion of the
materiality and form of maps and GIS. In this chapter I will therefore take a look
at these aspects of mapping. Perhaps more than elsewhere in the book I use the
term mapping very broadly, as anything that speaks to what Gaston Bachelard once
called the "poetics of space" (Bachelard 1958/1969).

   Bachelard's book, which is now a classic in architecture, planning, landscape design,
art, and other fields, sought to approach space not as container, box, or even by
its function but rather as a lived experience that encompassed daydreaming
(oneiric architecture), the everyday moment, and meaningfulness. Bachelard, who
first wrote the book in 1958 (it was translated into English in 1964), was a
phenomenologist and for most of his career a philosopher of science. His book has
more than a little flavor of Heidegger. For example, Bachelard places a great deal
of emphasis on "dwelling," which is not just to be in a particular place and time as
being present, but as beings with a rich past and future. In this way you could say
that he was aiming to avoid the same metaphysics of presence that characterizes
Heidegger's arguments, whereby mere presence is privileged, along with all its attributes
and exploitations, rather than the actual question of being and finding our place
in the world.

   Bachelard's approach to space is extremely homely, literally emphasizing the home
place in which we dwell; it is "our corner of the world," says Bachelard (1958/1969: 4).
We can dream and let the imagination roam in the house, where, safely sheltered,
it is the locus of our memories. Denis Wood and Robert Beck's *Home Rules* (Wood

and Beck 1994) makes a similar point. It is in your house that you emplace so many of your memories.

This idea is reminiscent of the art of memory of the Greek poet Simonides, who used the metaphor of the house as a mnemonic. Walking around a house (or the marketplace or other deeply familiar place) in the imagination, Simonides would place particular memories in loci such as alcoves, rooms, or steps. Later, he would reconstruct the walk and collect the memories again. This "method of loci" or memory palace was a well-known technique, taught throughout medieval times, and is described by the famous Roman orator Cicero.

Because of this deep link between memory and place, Bachelard coined a new form of analysis, topoanalysis, or "the systematic psychological study of the sites of our intimate lives" (1958/1969: 8). For Bachelard, this intimate space begins with the house.

In this it perhaps had something in common with Georges Perec whose "Species of Spaces" (*espèces d'espaces*) epitomized the "infra-ordinary" of our daily lives and banal habits, not as boring over-familiar routines but as in fact under-examined (Perec 1974/1997). It is these little events and practices of which our lives consist. This viewpoint has proven influential on a generation of geographers who take the everyday "psychogeographies" or the way we perceive space as their topic (Pinder 2003). Many of these people, such as Guy Debord who coined the term, the Boston art collective Glowlabs, and individual artists such as kanarinka (see kanarinka 2006) work in the intellectual vein opened up by psychogeographies of the 1960s (Wood 2007a).

It is interesting that so many authors have dwelt on the house because this is not a space usually associated with mapping or GIS. It is too small, too specific. Houses vary too much, there are also apartments to consider, condos, second homes, mountain cabins, dorm rooms, caravans, and trailers. And what is so important about a house?

But after all what is more important than your living space? For some years my neighbors owned a dog which they made live outside in the backyard. The dog was very fond of barking, which it would do all day and night. I can honestly say that it nearly drove me mad! (Luckily they moved out; if they hadn't I would have.) When your space and sanctity are violated it thrusts up into sharp relief many of the things you take for granted. We are material beings it says; we have to be somewhere, our bodies have to dwell and inhabit a particular space. When you take a new job you have to live in that place and so it matters very much what it's like.

Details count. That's the message. Perec's "Species" is disquietening and enlightening in that its perpetual listing off and merciless examination of the ordinary acts to defamiliarize you from your surroundings. Yet it's also incredibly witty because Perec picks out the kinds of things which, although you've never had that *precise* experience, are instantly recognizable:

In an old house on the 18th arrondissement I saw a WC that was shared by four tenants. The landlord refused to pay for the lighting of the said WC, and none of the four tenants was willing to pay for the three others, or had accepted the idea of a single meter and a bill divisible into four. So the WC was lit by four separate bulbs, each controlled by one of the four tenants. A single bulb burning night and day for ten years would have obviously been less expensive than installing one of these exclusive circuits. (Perec 1974/1997: 44–5)

"Species of Spaces" builds up a series of locales of ever-increasing scale, each of which occupies a section of this long essay: The Page, The Bed, The Bedroom, The Apartment, The Apartment Building, The Street, The Neighborhood, etc. (Perec hated the "etc." of lists like this, insisting that everything should be listed!) It's no coincidence that Perec began his spaces with "the page." Although it might seem that a page is not a space, but actually, Perec writes:

I write: I inhabit my sheet of paper, I invest it, I travel across it.
   I incite blanks, spaces (jumps in the meaning: discontinuities, transitions, changes of key).

I write
in the margin

I start a new
paragraph. I refer to a footnote*

I go to a new sheet of paper.

* I am very fond of footnotes at the bottom of the page, even if I don't have any-thing in particular to clarify there. (Perec 1974/1997: 11)

The page is a space then just like the others and Perec can inhabit it just as Bachelard inhabits the house, he can travel and type across it, or escape it into the margins. Next in the sequence is the bed, which like the page, we mostly inhabit longways. A page, a house, a map: space. Perec's novel *Life: A User's Manual*, takes this concept further. It tells the stories of the inhabitants of just one apartment building in Paris, but being a member of OuLiPo (society for potential literature) Perec traces it out as a knight's journey, moving from apartment to apartment as a chess knight would move. (Perec makes a mistake between move 65 and 66.) Perec thought of this structure of space as a way of generating ideas. He didn't want to be formulaic, but at the same time wanted to explore a certain kind of structure, one that was meaningful to his playful mind.

   A similar story is told by the American composer John Coolidge Adams, who wrote the opera *Nixon in China*. When he was younger he devised a piece that:

... was founded on the chance principles of John Cage, and was inspired by Henry David Thoreau's description of climbing Mt. Katahdin, in Maine. I obtained a map of the area and, with a compass, drew around the summit of the mountain a circle with a radius of fifty miles. This, I estimated, was the view that Thoreau described. I then made a list of all the local Abenaki place names – Millinocket, Ambajejus, Matawamkeag – and set these words to short, tonal melodies. (Adams 2008: 36)

In pursuing these practices the artist cedes a measure of control over the work, agreeing, once the principle of the concept is established, to abide by the results that it throws up. This in turn is reflective of the situation we often find ourselves in; dealing with outcomes from some initial choice that are not quite predictable and not quite random.

In 2000, a novel appeared which took these ideas and truly subverted them. Mark Z. Danielewski's *House of Leaves* is both a horror story and a spatial disruption. The novel is almost impossible to describe and has something of the flavor of a modern *Finnegans Wake*, but with wild typographic invention (some parts of the pages run in different directions or are printed backwards, some are almost empty, or have multi-layered footnotes; like Perec's *Life*, the novel has an index). Superficially it's about a family – the Navidson's – who move into a house on Ash Tree Lane. Will Navidson is a cinematographer, Karen his wife, and Chad and Daisy their children. One day they go away for four days to attend a wedding. When they return they find a "spatial violation" that is deeply uncanny or *unheimlich* in German (there follows a lengthy footnote in the novel into Heidegger's notion of uncanniness).

In their absence, the Navidson's home had become something else, and while not exactly sinister or even threatening, the change still destroyed any sense of security or well-being.

Upstairs, in the master bedroom we discover along with Will and Karen [on the video that is supposedly being taken] a plain, white door with a glass knob. It does not, however, open onto the children's room but into a space resembling a walk-in closet. . . . (Danielewski 2000: 28)

This space was not there previously, as photographs prove. The only conclusion is that it somehow appeared. Further investigation reveals that it leads to a long corridor, which after a while opens up into a huge room with a massive flight of steps going down, and although things become less clear here, it takes many hours to travel down this wide staircase. Danielewski's skill at creating a sense of horror at this spatial violation is immense, it is perhaps the only spatial horror story ever written. It is not so much the actual facts of the situation that manifest the horror, but rather a sense of dread often encountered in dreams. Fear is active; you scream, you run. But dread is passive, it drains and permeates you; it retilts your orientation to the world. It is the "affect" of a situation – the emotional penetration you feel. The safety of the house has been violated as we inhabit the paper (and of course there is a pun on house of leaves = book).

Perec, Bachelard, Debord, and the psychogeographers seek to evoke this affect or emotional response to the environment – although not necessarily of dread! Perec offers as a practical exercise the task of describing the street in *all* its ordinariness. This kind of exercise was later taken up by psychogeographers and map artists working to de/re-familiarize us with our environment and lived world: "You must set about it more slowly, almost stupidly. Force yourself to write down what is of no interest, what is most obvious, most common, most colourless" (Perec 1974/1997: 50)

Why on earth would anyone do this? Why not map out and record what is objective, important, not obvious? An answer is offered by de Certeau, who argues that a map is an act of destruction and forgetting. When we map something, especially in the cool contours of the grid, we are destroying what is there, namely the being in the world itself:

> Surveys of routes miss what was: the act itself of passing by. The operation of walking, wandering, or "window shopping," that is, the activity of passers-by, is transformed into points that draw a totalizing and reversible line on the map. (de Certeau 1984: 97)

For de Certeau, the map is an act of forgetting, contrary to Harley's position that it is an act of memory (see Chapter 7). Here we might also recall Monmonier's well-known line: "not only is it necessary to lie with maps, it is essential" (Monmonier 1991: 1), which puts a necessary lie (cf. Plato's noble lie) at the *essence* of maps. This apparent dispute can be resolved by agreeing with both authors, for the act of mapping is indeed one of creative destruction. Maps, as many critical geographers seem to fear (see Chapter 1), desubjectify and totalize; they also memorialize and create. Perhaps all acts of creation are ultimately also acts of destruction.

Psychogeography and Art-Machines

In 2004, psychogeographers and artists performed a piece called "Funerals for a Moment." Working from the idea that if these everyday moments are seriously that meaningful we should formally mark their passing with a small funeral, the group opened a website in which people could record something that happened (or not) as well as its exact time and location. For example, one entry dated Thursday April 22, 2004, 4:45 p.m., reads:

> As I was walking down Warren St. in the Cobble Hill area of Brooklyn, the wind suddenly picked up its pace. From this gust, I noticed an empty plastic bottle of soda begin to roll perfectly down the sidewalk. It rolled absolutely straight.

In the Funerals project, which was coordinated by the non-profit artist organization iKatun, a series of people with designated roles (usher, mourner, etc.) would return to the exact spot where the moment had taken place. A brief memorial would be read with a re-enactment of the event. Flowers would be thrown to mark the occasion (Figure 12.1).

Psychogeography draws initially from the surrealist and later the situationist movements of the twentieth century. Guy Debord, who is perhaps the most well known of the situationists, defined psychogeography in relation to other geography as follows:

Figure 12.1 Scenes from "Funerals for a Moment" by kanarinka 2004. Photos by Joshua Weiner. Used with permission.

> *Psychogeography* could set for itself the study of the precise laws and specific effects of the geographical environment, consciously organized or not, on the emotions and behavior of individuals. The adjective *psychogeographical*, retaining a rather pleasing vagueness, can thus be applied to the findings arrived at by this type of investigation, to their influence on human feelings, and even more generally to any situation or conduct that seems to reflect the same spirit of discovery. (http://library.nothingness.org/articles/SI/en/display/2)

For Debord then, the very development of statistical and thematic mapping that had established itself in the nineteenth century (discussed in Chapter 6) was actually counter-productive.

Denis Wood has described the work of this time by drawing a parallel between the Situationists in Paris and the work of urban planner Kevin Lynch and others at Clark University in the 1960s (Wood 2007a). Methodologically, the Situationists, interested as they were primarily in the urban environment, employed the *dérive* (the "drift"). The *dérive* was usually undertaken in a small group, according to Debord, two or three people were optimal because this allowed cross-checking and the possibility of an objective conclusion. This was not a random walk, but was influenced by the contours of the "psychogeographical relief" and the Situationists prepared a number of maps of Paris in the 1950s to show the flows and forces they perceived in the environment.

The recent independent film *Adrift in Manhattan*, directed by Alfredo de Villa (2007), evokes this sense of the infra-ordinary. Its three main characters journey to work, their paths, often unknowingly, intersecting. They are "adrift" not only in the sense of the *dérive*, in the sense that you cast a boat adrift, so that it floats around – neither randomly nor quite predictably, but also in the sense that they sometimes are disconnected from their emotions. Heather Graham plays Rose, an optometrist in Chelsea who has lost her two-year-old son in a terrible accident and she and her husband are now separated. Meanwhile Simon (Victor Rasuk), a young photographer who barely speaks, takes pictures of her in the park and follows her in a way that almost or probably is stalking. As his way of communicating Simon shoves some pictures he's developed (he works in a camera store) through her letterbox, after making sure that the store's name is one back. When she goes to the store seeking answers she meets Simon unexpectedly and they seem to at least tenuously connect. He: fascinated by her scarf. She: strangely attracted to this naïf. One of Rose's clients is Tommaso (Dominic Chianese), a painter who is suffering from fading eyesight. Sight, seeing, and surveillance are major themes here. So all three are connected (in one subtle scene we learn that Tommaso and Simon share the same apartment building). A lot of the movie takes place on the subway or the streets of the city as the lives of the characters, their friends, and family intersect. The ending does not resolve or close these storylines; everything remains open, though we are no longer watching.

If the art practices of the Dadaists and the Situationists left a legacy, it was that art – and map art – was not limited to the traditional painting, but could be a practice. Over the last few years artists have gathered with increasing frequency at events that foreground practice, inspired by New York-based Glowlab's psy-geo-Conflux in 2003.

One of these artists, kanarinka, begins from the premise that "Any representation of the world that asserts its neutrality and objectivity is immediately suspect" (kanarinka 2006: 25). She continues:

> The question now for artists (and likely for cartographers) is emphatically not how to make a "better" picture of more "accurate" map. The world in fact, needs no new representations at all. It needs new relations and new uses: in other words, it needs new events, inventions, actions, activities, experiments, interventions, infiltrations, ceremonies, situations, episodes and catastrophes. We have departed from a world of forms and objects and entered a world of relations and events. But we still desperately need art and maps. Is it possible to think of a map not as a representation of reality but as a tool to produce reality? (kanarinka 2006: 25).

Here kanarinka is close to the ideas of psychogeographers, but also those of the philosopher Alain Badiou. For Badiou the very question of being is bound up with the event or happening. For Badiou, events leave "traces," they change things and "leave a cut in the world." Artistic events in particular have this transformative action, and between the trace they leave behind and the relation of the body to the world there is the subject. The body is either characterized as entirely material, says Badiou (the subject is reduced to the body, your identity is defined by your body), or the subject is seen as entirely transcending the body – dualism. Neither of these will do, so Badiou's suggestion is that we need a third subjectivity which does not reduce the subject to the trace of the event nor the body in the world. The relational difference between event and the world is the space in which the subject is constituted.

An event then is an affirmative intervention. Of course it has consequences, and in later lectures Badiou has discussed the relation between politics and art, as well as the responsibility of artistic creation to find new forms of subjectivity.

Recovering Art

One of the figures most primarily associated with work on art and mapping is the geographer Denis Cosgrove (1948–2008). Like many of the authors mentioned above, Cosgrove placed emphasis on the act of mapping itself as a site of imagination. "Acts of mapping are creative, sometimes anxious, moments in the coming to knowledge of the world," he once wrote (Cosgrove 1999: 2). Cosgrove also wished to recover the junctions between art and mapping that apparently seemed to go their separate ways during the twentieth century. In a recent summary paper (Cosgrove 2005), he argued that the art–science difference was more apparent than real, and discussed numerous examples of artistic mapping expressions (Cosgrove 2006; Cosgrove and Della Dora 2005). If Woodward's *Art and Cartography* (Woodward 1987) had challenged the notion that with the incorporation of surveys and other scientific impedimenta this meant that cartography had progressed from art to science, Cosgrove in turn challenged Woodward's association of art with aesthetics. In other words, art was not about how pretty something is.

For example, this late nineteenth-century map epitomizes the heights of baroque detail around the margins of the map (see Figure 12.2). Britannia herself sits astride the whole world at the center of the map, which in turn is held up by Atlas. Britannia is dressed as a Greek goddess herself, with her characteristic classical helmet pushed back from her brow (just as modern American football players do today). She forms the center of a tableau of offerings, dark-skinned and naked natives gaze up her majesty, one offers a cornucopia, another a fantastic peacock's tail. Representatives of the British Empire pay homage; even Nature faces toward her, its natural instincts calmed in her presence; a tiger ignores the live birds and submits to be held on a chain by an Indian officer. The map itself shows the extent of the British Empire in red, just as it should be. The map is cast on a Mercator projection with Great Britain in the center of the map on the newly agreed (1884) international zero meridian at Greenwich.

In fact this was a fairly late map (1886), which was made to support the arguments of a nascent Imperial Federation which would supersede the empire. As Biltcliffe has recently shown, this map is doubly interesting because the artist, Walter Crane,

Figure 12.2   *Imperial Federation Map of the World Showing the Extent of the British Empire in 1886.* Photo courtesy of Newberry Library. Used with Permission.

was a socialist who was a leading member of the Arts and Crafts Movement and a student and follower of John Ruskin (Biltcliffe 2005). Long before Walter Benjamin more famously denounced the effects on art of mass production, Crane strove to work against the ugliness, as he saw it, of Victorian consumer society.

> Crane believed that society's contemporary emphasis on commerce and capitalism, the foundation of the imperial project, enslaved humanity to an economic system that produced all things for profit rather than use and destroyed the vitality of the people. (Biltcliffe 2005: 65)

Although socialism opposed imperialism, Crane did believe that empire could be used to advance social reform. If you look more closely at the map, you can see that Atlas is labeled "human labour" (see detail, Figure 12.3). The worker's strength then, is what holds everything up, including empire. Were Crane's illustrations then secretly subversive of the apparent message of the map? Were there multiple plays of power across this map, rather than the simple one of imperialism? Whatever the answers to these questions it goes to show that map art is not mere decoration or aesthetics.

For Cosgrove, the claim that the marginalia of the map is aesthetics as opposed to the analytical rigor of science acts to undermine the very role of the visual in

Figure 12.3 Detail, *Imperial Federation Map of the World Showing the Extent of the British Empire in 1886.* Photo Courtesy of the Newberry Library. Used with Permission.

modern science. Science relies upon the visual persuasiveness of its images, including maps. Using Bruno Latour's concept of the "immutable mobile" which he glosses as "an instrument that preserves the meaning and truth claims of scientific observations as they circulate across space and time," Cosgrove suggests that the scientific map has to transform its information into a universal language, just as much as does the artistic production of spatial images. The twentieth century's claims for a modern, scientific status of mapping act only to obscure the "shared epistemological status of art and science" (Cosgrove 2005: 37).

For example, the Irish artist Deirdre Kelly who now lives and works in Italy, works with mixed media and collage to produce a number of works that draw on cartographic imagery. In her piece "Credit" (Figure 12.4), commissioned by Reuters Ltd, she depicted the global flows of capital as if on a dual hemisphere projection. In the lower part of the piece the moneylenders and financiers perform their deals, while in the upper part the Garden of Eden remains tantalizingly close. Old-fashioned scrollwork belies the modern messages inscribed on them; "Rollover," "Distress," "Risk." The extension and assessment of credit is here a blatant immutable mobile.

Figure 12.4 Top: "Credit" by Deirdre Kelly. Bottom: "Credit" detail. Source: Deirdre Kelly used with permission of the artist. Color original.

Figure 12.4 (*Continued*)

No wonder then that artists have been experimenting for more than a hundred years with ways of depicting space. These experiments go at a orthogonal angle from the academic disciplinary history of mapping. Where for the latter increasing accuracy and verisimilitude are the goal, for map artists the very practice of representation became a stake to be problematized.

If the avant-garde movement is often dated to the Salon des Refusés of 1863, which featured artworks that were rejected (*refusés*) from the main exhibition of the Académie des Beaux-Arts of the Paris Salon, it was perhaps with the impressionists that the question of representation was explicitly raised. Claude Monet's eponymous work *Impression, soleil levant* (*Impression, Sunrise*, 1872) has characteristically visible brush strokes and vivid colorings rather than the smoothly blended brush strokes more typical up to that point. Monet, Renoir, and others organized an association for exhibiting art independently in 1873, and its members were encouraged not to take part in the official Salon. With Paul Cézanne, Edgar Degas, Mary Cassatt, and Monet these aspects make the viewer pause as the artwork brings into question what it means to represent the world. Noteworthy too is their oppositional stance.

In the following decades artists would explicitly take this question up in many different ways and media, and so form the artistic movement we know as "modern art" which includes the avant-garde, impressionism, Dadaism, and installations. If the year 1900 could see the hosting of a major exhibition in Paris as part of the "Exposition Universelle," to welcome the new century, most of its art would not have looked out of place a hundred years before (P. Wood 2003). The main traditions of art going back to the Renaissance – history-painting, portraitures, nudes, landscapes, still lifes, and sculptures – dominated the exposition. But just as cartography was grounding itself more solidly in the scientific method, artists, many of them working in geographical and spatial themes, came increasingly to challenge what representation itself was.

Although they took different routes to explore this question, one thing they were interested in was whether realistic and naturalistic art had been wrong all this time in trying to show the landscape as realistically as possible. Perhaps there were deeper truths to be had by reconfiguring space? At first sight, for example, Georges Braque's *Clarinet and Bottle of Rum on a Mantlepiece* (1911) appears hopelessly confused and complicated. While further inspection does allow you to pick out the clarinet (horizontal, about midway up), and possibly some music scrollwork (or is it a cornice of the mantelpiece?) in lower right, the overall impression is of a confusion and overlapping of space and spatial arrangement. The piece is suggestive more than definitive and permits multiple interpretations – perhaps we could say it shows *some* truths rather than *the* truth.

Braque's piece is also remarkable because it contains lettering and writing. The "rhu" of the French word for rum is visible, as well as "valse" (waltz), which articulates music neither as sound nor picture but as text, an idea echoed in the bass notes and clefs. Finally, Braque brazenly adds in the nail and its shadow upon which the painting hangs, in case we should forget that this is a painting. To Braque therefore goes the honor of the most famous nail in all artwork!

More recently the British artist Tony Cragg has worked with plastic garbage that has washed up on beaches. In his *Britain Seen from the North* (1981), the piece can either be seen from a distance, where it resembles a sideways map of Great Britain facing down a figure who may be the artist (or the viewer), or it can be seen closer up, where the overall image dissolves into the pieces of plastic rubbish that comprise it. Although Cragg is British he has lived in Germany since the late 1970s and the piece may reflect his status as an outsider, or the difficulties confronting the country at the time (the Conservative government of Margaret Thatcher was engaged in a long and bitter campaign to suppress the unions). Cragg's work also serves to question the London-centrism of the country, as it entered the post-punk era of the late 1970s and the wedding of Prince Charles and the nationalism surrounding that. In an artist's statement he has written:

Cutting up material, turning it round, changing the contours, the surfaces and the volumes time and time again. . . . Making sculpture involves not only changing the form and the meaning of the material but also, oneself. Your feelings and your thoughts about what you see change constantly. The popular and unhelpfully simplifying dichotomies of form and content, ugly and beautiful, of abstract and figurative, expressive and conceptual dissolve into a free solution, out of which a new form with a new meaning can crystallise. (Cragg n.d.)

Conclusion

When did map art begin? Wood has recently argued it began just after World War I (Wood 2008). You could, if pushed, admit that map art in the sense we know it today developed through the 1920s with surrealist works (the 1929 surrealist map of the world, see Figure 2.3), and then through a number of individual works (Max Ernst, Joseph Cornell in the 1930s), Marcel Duchamp in the 1940s, then perhaps the letterists and psychogeographers (Guy Debord and his colleague Asger Jorn). Jasper Johns in the 1960s had several well-known pieces called simply *Map*. Since the 1950s, the Argentinian-Italian artist Lucio Fontana has been working on a whole range of artistic works including sculptures, slashed paintings, and installations under the rubric "Concetto Spaziale" (or "Spazialismo," spatialism). Fontana drew from both the cubist and futurist movements, and his 1946 "manifesto" outlined a way of getting beyond the painting into what he called a "free space." The characteristic slashes and cuts of Fontana's spatialist pieces create a space that is neither that of the surface of the painting, nor yet that which it is representing (the peri-space around it), but a third indeterminable space. Like Braque, it also highlights and problematizes the act of representation itself.

How can we assess all of these people, these artistic movements? For many people aesthetics is the meaning of art, and those who produce it the best are artistic geniuses. While I would not wish to deny the aesthetic experience of art, this is an inadequate understanding of map art (and perhaps of all art) by itself. A more adequate understanding would recognize at least three other components of map art: that it brings to light other people's worlds; that it produces or articulates a shared understanding of our world; and that it can reconfigure cultures into something else. In other words rather than fundamentally representation or aesthetics, art is concerned with questions of truth.[1]

We can conclude with a few examples offered by the Boston-based artist Catherine D'Ignazio (a.k.a. kanarinka). She identifies three main groupings of map-art activities that followed from a "spatial turn" of the arts over the last decade or so (kanarinka 2009). D'Ignazio suggests that with the acceleration of modern life and space–time compression we have called forth new ways of representing the world to ourselves:

> The accelerated accumulation and circulation of capital, conflict, and people around the globe is a phenomenon that required (and is still requiring) diverse societies to develop visual and cultural mechanisms for articulating their relationship to the "whole" world, a world which, economically and technologically speaking, is already right in their backyard. (kanarinka 2009)

In this she follows arguments that David Harvey made in his book *The Condition of Postmodernity* (the phrase "how we represent the world to ourselves" is his, see Harvey 1990: 240). Expanding on Harvey's influential book, she offers three categories of map art that characterize contemporary map-art practices: Symbol Saboteurs, Agents and Actors, and Invisible Data-Mappers.

Symbol Saboteurs are "artists who use the visual iconography of the map to reference personal, fictional, utopian or metaphorical places." Thus the contours of maps are here reappropriated for the artist's own purposes. Nina Katchadourian's cutout maps removed everything from maps of *Austria* (1997) except the roads, which were then tangled together in balls, like long strips of DNA. The work of Jasper Johns and Tony Cragg mentioned above would also illustrate this grouping.

Working in the spirit of humor and irony, Catherine Reeves created the so-called "Equinational Projection" to make the most politically correct map possible following the great map controversy over the Peters projection. In her new projection, originally published in an obscure fanzine, Reeves allocates the exact same size and shape for every country on earth (Figure 12.5). Neither Peters advocates nor Peters denialists can object!

The second grouping D'Ignazio calls Agents and Actors, "artists who make maps or engage in situated, locational activities in order to challenge the status quo or change the world." After World War I, for example, the artistic movement known as Dadaism reacted against the chaos, loss of life, and redrawing of boundaries by working in a number of avant-garde projects aimed at critique or social change. Max Ernst's *Europe after the Rain I* (1933) is based on a reconfigured map of Europe, the well-known *Surrealist Map of the World* (1929, anonymous, but attributed to Paul Eluard by Denis Wood [Wood 2007b], see Figure 2.3), played with geography

IASBS EQUINATIONAL PROJECTION

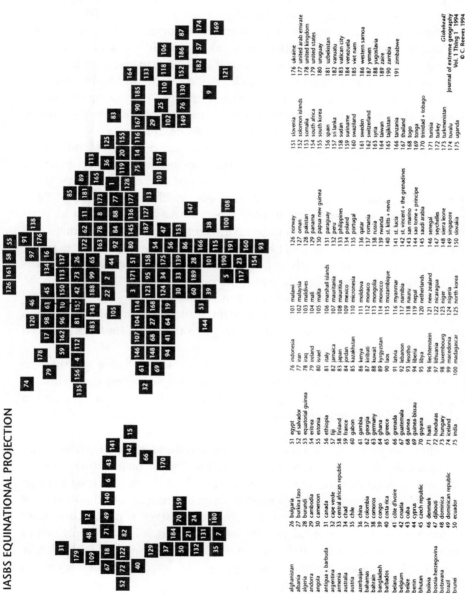

1 afghanistan
2 albania
3 algeria
4 andorra
5 angola
6 antigua + barbuda
7 argentina
8 armenia
9 australia
10 austria
11 azerbaijan
12 bahamas
13 bahrain
14 bangladesh
15 barbados
16 belarus
17 belgium
18 belize
19 benin
20 bhutan
21 bolivia
22 bosnia-herzegovina
23 botswana
24 brazil
25 brunei

26 bulgaria
27 burkina faso
28 burundi
29 cambodia
30 cameroon
31 canada
32 cape verde
33 central african republic
34 chad
35 chile
36 china
37 colombia
38 comoros
39 congo
40 costa rica
41 côte d'ivoire
42 croatia
43 cuba
44 cyprus
45 czech republic
46 denmark
47 djibouti
48 dominica
49 dominican republic
50 ecuador

51 egypt
52 el salvador
53 equatorial guinea
54 eritrea
55 estonia
56 ethiopia
57 fiji
58 finland
59 france
60 gabon
61 gambia
62 georgia
63 germany
64 ghana
65 greece
66 grenada
67 guatemala
68 guinea
69 guinea-bissau
70 guyana
71 haiti
72 honduras
73 hungary
74 iceland
75 india

76 indonesia
77 iran
78 iraq
79 ireland
80 israel
81 italy
82 jamaica
83 japan
84 jordan
85 kazakhstan
86 kenya
87 kiribati
88 kuwait
89 kyrgyzstan
90 laos
91 latvia
92 lebanon
93 lesotho
94 liberia
95 libya
96 liechtenstein
97 lithuania
98 luxembourg
99 macedonia
100 madagascar

101 malawi
102 malaysia
103 maldives
104 mali
105 malta
106 marshall islands
107 mauritania
108 mauritius
109 mexico
110 micronesia
111 moldova
112 monaco
113 mongolia
114 morocco
115 mozambique
116 myanmar
117 namibia
118 nauru
119 nepal
120 netherlands
121 new zealand
122 nicaragua
123 niger
124 nigeria
125 north korea

126 norway
127 oman
128 pakistan
129 panama
130 papua new guinea
131 paraguay
132 peru
133 philippines
134 poland
135 portugal
136 qatar
137 romania
138 russia
139 rwanda
140 st. kitts + nevis
141 st. lucia
142 st. vincent + the grenadines
143 san marino
144 sao tome + principe
145 saudi arabia
146 senegal
147 seychelles
148 sierra leone
149 singapore
150 slovakia

151 slovenia
152 solomon islands
153 somalia
154 south africa
155 south korea
156 spain
157 sri lanka
158 sudan
159 suriname
160 swaziland
161 sweden
162 switzerland
163 syria
164 taiwan
165 tajikistan
166 tanzania
167 thailand
168 togo
169 tonga
170 trinidad + tobago
171 tunisia
172 turkey
173 turkmenistan
174 tuvalu
175 uganda

176 ukraine
177 united arab emirate
178 united kingdom
179 united states
180 uruguay
181 uzbekistan
182 vanuatu
183 vatican city
184 venezuela
185 viet nam
186 western samoa
187 yemen
188 yugoslavia
189 zaire
190 zambia
191 zimbabwe

Globehead!
Journal of extreme geography
Vol. 1 Thing 1 1994
© C. Reeves 1994

Figure 12.5 Catherine Reeves, "The Equinational Projection." Source: *Globehead!* 1, 1994, pp. 18–19. Used with Permission.

by downgrading or omitting the superpowers and enlarging Easter Island. More recent work by a number of artists has used cartographic-like symbology to critique more recent military engagements in Iraq and Columbia, such as Lize Mogul and Dario Azzellini's *The Privatization of War* (2007), or Trevor Paglen's map of extraordinary rendition flights in *Torture Taxi* (Paglen and Thompson 2006).

The Fluxus art movement, which emphasized performance art and "events" (cf. Badiou) in the 1960s–70s, questioned the value of traditional art (as well as its high-minded seriousness). Following Yoko Ono's 1962 call to create an imaginary map and use it to navigate a real city, Anna María Bogadóttir and Malene Rørdam made *New Copen York Hagen* (2004), in which a map of Copenhagen was superimposed on New York City and used as a tour guide. In this category we might also place the Situationist International and psychogeography.

Finally the Invisible Data-Mappers are "artists who use cartographic metaphors to visualize informational territories such as the stock market, the internet, or the human genome." Here it is data itself that is mapped and centered as an issue. Kelly's *Credit* discussed above is one such example, which traces the global flows of borrowing, risk assessment, and interest. The Internet has been often mapped as we saw in Chapter 10 and the boundary between these informational maps and map art is not always clear. In *1:1* (1999) Lisa Jevbratt investigates the idea of the 1:1 map discussed by writers such as Lewis Carroll, Umberto Eco, and Jorge Luis Borges by visualizing the entire data set of Internet Protocol (IP) addresses on the Internet. These possible IP addresses are then color-coded according to whether they are actually taken up by a website. Martin Wattenberg and Marek Walcak's installation *Noplace* (2007–8) is a multimedia piece that takes information feeds to produce visions of utopia.

John Luther Adams' *The Place Where You Go to Listen* (installed in the Museum of the North in Fairbanks, Alaska) is a sound and vision experience, which takes geological, geothermal, and atmospheric readings of the earth to produce an ever-changing series of sounds and images. Earthquakes, clouds, and even the moon register in the soundscapes produced.

These three categories may be a way to think through the map art of the last century or so, but equally they may also be transient and ultimately unable to capture the diversity of mappings that are now being produced. There is clearly little danger that art in the early twenty-first century is indistinguishable from art of a hundred years ago, as was the case for the Paris exposition. And yet if Harvey, D'Ignazio, and other observers of art and culture are to be believed, today's art is just as deeply tied to contemporary society, just as ardently critiquing it, making fun of it, endorsing or opposing it. It is significant that maps and mapping are so much a part of that project.

Note

1   Those interested in taking this up further may wish to consult Heidegger (1993) and Dreyfus (2005).

Chapter 13

Epilogue: Beyond the Cartographic Anxiety?

But then, what is philosophy today – philosophical activity, I mean – if it is not the critical work that thought brings to bear on itself? In what does it consist, if not in the endeavor to know how and to what extent it might be possible to think differently, instead of legitimating what is already known?

(Foucault 1985: 8–9)

I began this book by pointing to the conflicted position in which maps and GIS seem to be held today. On the one hand maps seem as popular as ever and there are few people working in geography, the environment, ecology, archaeology, paleontology, geology, and planning (the "geo"-sciences) who could imagine doing without maps or GIS. On the other hand, maps and GIS do seem to have this "perverse sense of the unseemly" to use John Pickles' felicitous words quoted earlier. They seem unreconstructed, doomed by "ontologies," heavily complicit in imperial/colonial power, and no doubt ultimately the concern of map librarians or GIS "techies." In sum: the cartographic anxiety.

Over the years I have encountered my fair share of both these attitudes (even sometimes held by the same person). In writing a book for a series called "Critical Geographies" as this book was, I have endeavored to do three things in response. The first is to explore what possibilities there may be for a critical cartography and GIS. This has meant asking the question, what does "actually existing" critical cartography and GIS look like? Who has been doing it? A significant number of people simply have no concept of critical GIS or cartography. So the first step, which occupies roughly the first half of the book, is to explore what critical cartography and GIS might be. This has meant answering the question "what is critique?" through examples, but also looking historically at the emergence of modern GIS and cartography.

This historical lens is an important one. If history is our collective memory we like to tell ourselves (with the naughty bits and reverses edited out), it is therefore also the site of the greatest significance. (If you doubt this, ask yourself why in Orwell's

Nineteen Eighty-Four the regime employs Winston Smith to edit and glorify not the *future*, but the *past*.) From our sense of history we draw our sense of ourselves, and for a Heidegger or Marx, we are born into this inherited world. It is one of the contributions of critiques then that they recover those subjugated knowledges that do not fit with the established categories of knowledge. If for example, the standard textbook account of modern cartography/GIS is that it became scientific following the work of Arthur Robinson in the post-war era, why has there been so little written about his time at the OSS and the way he denatured the discipline? Or again, if Peters denialists wish to invoke James Gall as the originary inventor of the Peters world map, what does it say that they are appealing to a Preadamite who explained away fossil remains by attributing them to a now extinct race of angels descended from Satan? Simple stories of heroes and villains will not do.

The second step is to put this critique into practice. Therefore this book is written, as best as I can, in a critical spirit. But from whence is it launched? Here I have to acknowledge my own positionality as someone who "does" critical cartography and GIS. But this "here" is hard to locate; I exist between two worlds in a kind of exile (Said 2000), not only because of my British background but because I have lived in America for over 20 years. As Said experienced, there is still a sense of alienation from one's adopted country that cannot be shaken off. But this is doubled, or mirrored, by the fact that I also sit somewhat uneasily between the two worlds of critical social theory and mapping/GIS. There is a very real danger and actuality of being dismissed by both sides, my theoretic qualifications are never sufficient for critical geographers, but are too sufficient for GIS users. This is not to claim an obscure "outsider" status; obviously I occupy a specific and privileged intellectual position. But neither is it truly an insider status; for many academics, writers, poets, and artists their degree of freedom is highly constrained and surveilled, and not just in other countries but also in the United States. What I experience is something like a voice from the edges, a question of belonging. Edward Said in his lifetime was at various times accused of being both too close to the Palestinian cause and of not being "authentically" Palestinian. As is perhaps the case for many people, I retain both a proximity to and a distance from the subject under discussion. And I think it is the same for a practice of critique.

So in this book I have tried to act as a kind of translator, believing that this would be most appealing to readers who find themselves, not quite comfortably, in one or more camps. This kind of "shuttle diplomacy" might act as a way of bringing thought to bear on itself, and of learning to think differently. Every translation is after all an invitation for a re-translation. Like many students I encounter, you might find that you occupy several positions at once in Figure 1.1. Perhaps you are attracted to the possibility of acquiring a recognized GIS Certificate, but also the possibilities of bottom-up user-produced maps. The clash of motives here may usefully spark that questioning at the heart of critical practice.

If we recall the three principles of critical geography outlined earlier, that is, it is oppositional, it is activist/practical, and it is embedded in critical theory, then a number of chapters attempt to put these principles into practice. For this book I chose three topics which seem to me to be important, and in which GIS and cartography play significant and problematic roles. These are governing with maps (Chapter 6), geosurveillance (Chapter 9), and the construction of race (Chapter 11). Other chapters, speaking to other issues, are possible and even necessary. That is the flaw in this book or any book that is also the opening for further seeking.

If we agree with David Harvey that "cartography is a major structural pillar of all forms of geographical knowledge" (discussed below) then one of the first things we will want to know is the nature of the forms of knowledge produced for these three domains. A critical approach will also want to problematize these areas, to work through their implications, contradictions, assumptions, historicities, and deployments.

Thirdly, I have tried to situate both the source and the target of critical cartography and GIS. Any account which tells the story from a purely disciplinary perspective will, it seems to me, omit some of the most interesting and radical practices of that critique. The fact is that mapping today is escaping the discipline. The rise of "people-powered mapping" and the geoweb at the same time that we have seen the rise of the political "netroots" and people-powered politics is not coincidental. They stem from the same cause and desire to create alternative forms of expression beyond those encompassed by the traditional power-holders (whether the geographical knowledge elites or Big Media). Critical cartography and GIS then, does an end-run around the accreted power structures such as academic experts, textbooks, and official "bodies of knowledge." In Chapter 12 for example, I tried to trace some of these non-disciplinary and non-academic critiques in the context of map art and the poetics of space.

In order to achieve its goals must critique place itself on the "outside?" If much critique is reflexive and internal the degree to which one can oppose from within is not an uncontroversial one. For some, opposition can only take place from the outside, from a purer position, detached and uncorrupted. Work within the system will only lead to becoming a part of the system, becoming co-opted. For others this very claim for detachment, of escape from the object of analysis, is only another sign of the impossibility of escape from the power relations of mapping and GIS. Again this is a question of positionality, and you may find yourself both within and outside the system at different times. I know I do.

The Necessity of Questioning

If critique has achieved anything in GIS and cartography it has been to intervene, whether from outside or within, and to challenge the assumptions of the production of geographical knowledges. This is a spirit of questioning in which we ask what other ways of doing or thinking are available? If critique is not simply absorbed by "disciplinary" cartography and GIS in a safe, enclosed manner ("the response of the oyster"),[1] then at times it has established a pull in an altogether different direction.

This view was recently put forward by David Harvey, who identified a number of "sites" for the production of geographical knowledge – cartography among them. Harvey wondered why this site of production had been so overlooked:

> cartography is, plainly, a major structural pillar of all forms of geographical knowledge . . . it is years now since Foucault taught us that knowledge/power/institutions lock together in particular modes of governmentality, yet few have cared to turn that spotlight upon the discipline of Geography itself. (Harvey 2001: 217, 220)

And as Livingstone has recently shown this knowledge can be produced in very real and geographic sites, often with a specific cartographic contribution (Livingstone 2003).

One of the world's oldest religions – Judaism – every year marks the great tradition of the Passover Seder to remember and reflect on their exodus from slavery. In the tradition, great emphasis is placed on questioning and grasping this story, as the family and friends gather together around the table. How was this freedom achieved and of what did it consist? As the evening progresses, the story is often told of another gathering, in which four sons, sitting at the table, come to terms with Passover. Each son approaches the question of freedom differently (the wise son, the disruptive son, the simple son, etc.). For example, the wise son asks "what is the meaning of this ceremony?" By asking a good question, the wise son learns the most. There is also the disruptive son, who challenges the meaning and purpose of the whole thing, perhaps hoping to undermine it. Yet his spirit of questioning is strong. The simple son's questions perhaps operate at a different level.

Yet out of all these different responses, worst of all is the fourth son, who does not ask questions. This is worse than disruptive questioning, or even of *not having answers*. For this last son, the one who does not question, a change must be brought about, not externally, but within the son himself. Who must bring it about? Ultimately the son himself. The son must be brought to ask the question (symbolic foods may be placed on the table such as bitter herbs to evoke the bitterness of slavery). The same is true for those in slavery – how could they escape from bondage? Not until they brought about a change within themselves, that allowed them to question and conceive of other ways of being.

This thematic, of the relation between freedom and questioning, has many parallels. For an Enlightenment Kant it appears in his famous axiom *sapere aude!* (dare to know, or more literally dare to taste or sense: we are *homo sapiens*; the ones that are wise i.e., those that taste). Freedom is not a state – it's a process, or even a spirit, a style. Returning to George Orwell's *Nineteen Eighty-Four* we see the totalitarian state seeks a reversal of this in its slogan: "Freedom Is Slavery." Such a state will suppress all free thought by making it seem as if that's what freedom is: total obedience. By contrast, the protagonist Winston Smith asserts "[f]reedom is the freedom to say that two plus two make four. If that is granted, all else follows" (Orwell 2003: 83). Smith takes a stand (which indeed later proves dangerous) by asserting an obvious, but politically difficult, truth. All else follows from this, even freedom. Questioning doesn't have to be difficult (you can't get more basic than asking "what is two plus two?"), but it does have to occur.

For Michel Foucault, this practice of questioning exemplified the very practice of freedom, and again it came from a change or attitude from within. In one of his last interviews before he died in 1984, Foucault talks about the relationship between three key components: ethics, self, and freedom (Foucault 1997a). Freedom, he asserts, is a practice, not a place where one can get to, or a condition that one can finally achieve. The implication is that if freedom is not practiced through continued questioning, freedom will be lost. This is just what Big Brother wishes to assert by tricking people into accepting that freedom is slavery – the unquestioning acceptance of one's place.

"The concern for the self" sounds narcissistic and we might be tempted to overlay it with New Age interpretations (self-actualization, or self-knowledge), or a selfishness that puts oneself before the concerns of others, or a religious, pedagogical, medical, or psychiatric practice. Avoiding these, Foucault claims that you can recover a meaning of ethics as a "way of being" from the Greeks (Foucault 1997a: 286). This way of being was not a state of enlightenment but an attitude; a questioning:

> I believe that this is, in fact, the hinge point of ethical concerns and the political struggle for respect of rights, *of critical thought* against abusive techniques of government and research in ethics that seeks to ground individual freedom. (Foucault 1997a: 299, emphasis added)

The Seder, George Orwell, Immanuel Kant, Foucault – what brings these disparate subjects together is their emphasis on freedom as a practice of questioning that has the name critique. Even if sometimes we do not have a ready answer, we are still better off – we are free – to the extent that we do the "critical work that thought brings to bear on itself."

So where does this actually leave us? In Chapter 1 we envisaged mapping as a field of tension that is riven by a number of forces running through it in different directions (often at the same time). Like a spider's web, the field trembles and registers the impact of these forces, and reshapes itself from time to time (Figure 1.1). I'd like to conclude by suggesting that these different directions are a series of questions: where are we, and where do we go next?

One striking tension (question) lies in what at first sight looks like a technological issue, but which actually strikes deep to the core of cartography and GIS. Will we secure knowledge by and for a clerisy of "geographical knowledge elites" or will we establish more bottom-up production of geographical knowledge through Goodchild's citizen sensors? (Again, I remind readers that the answer may not be one or the other but both – or neither.)

For example, there are powerful institutional forces that are seeking to encapsulate cartography and GIS within professional disciplinary knowledges. In America the GIS Certification Institute offers applicants the nationally recognized certification as a GIS Professional (GISP). To date, over 4,600 GISPs have been awarded and are maintained on a central registry that can be consulted by employers. The idea then is that the certification demonstrates that the holder has mastered a core set of knowledge or GIS practices and is qualified as an authority or expert. Similarly, the recently published GIS "Body of Knowledge" (BoK), upon which the GISP certification explicitly draws, is a codification of the skills and issues which a GIS expert shall be expected to master (DiBiase et al. 2006). Notably, this BoK was initiated and published by the Association of American Geographers (AAG), the leading American disciplinary organization in geography.

As we have seen throughout this book, assertions of knowledge are questions of power relations. An extensible semantic web of "ontologies" for the National Map (Chapter 8) might be defended by stating that "if other people want to add their own, local descriptors they may do so," as was recently observed at a Specialist Meeting on the map. What this ignores is that "official knowledges" have an unseemly perversity of establishing themselves as the more important, vital, or necessary at the cost of local or indigenous knowledges. We have seen throughout this book that a politics of knowledge is never too far away from our endeavors.

Additionally, these assertions of knowledge are contested and contestable (counter-knowledges or counter-mapping). In 2008–9 for example, a major controversy began to surround the work of several American geographers working in Mexico, who were engaged in a community mapping project. At the heart of the controversy was the issue of whether these geographers, including the President of the American Geographical Society (AGS), had clearly identified the fact that they had accepted US Department of Defense funding to carry out their work. There were questions over the degree to which the geographers had obtained the permission of local indigenous groups, or of all the groups and whether that permission had been truly given freely. This was a particularly painful episode for myself and many other geographers because the motives of the American geographers initially appeared admirable; that is, technology transfer and self-empowerment of the indigenous groups, who lived in an area historically subject to repression, exploitation of the Mexican government, and of USA imperialism. Yet undoubtedly my colleagues showed fatal naivety, perhaps not only in accepting the military funding, but also in calling this work a "Bowman Expedition" after Isaiah Bowman, a previous president of the AGS and Johns Hopkins University. As Neil Smith has shown, Bowman's involvement in imperial geopolitics throughout the first half of the twentieth century is problematic (Smith 2003). (Smith's account of Bowman's racism is less convincing.)

This example shows again the necessity of what I called in Chapter 8 an existential or anthropological GIS (or mapping): one that can elucidate the cross-cultural ways of being (worlds). Assuming for a moment that my colleagues did declare their funding and gain their IRB approvals correctly, what we see here is a clash of worldviews. By Western standards of science they did nothing wrong, but it turns out that these standards are insufficient or just irrelevant for the indigenous peoples in Oaxaca.

If modern mapping and GIS came of age in the post-war period, they did so at a time when geographical enquiry was dominated by spatial science. As is well known, this quantitative revolution dominated intellectual activity in Anglo-American geography for a number of decades during and after the war, and its cartographies were deliberately blank and placeless. Bunge's manifesto (Bunge 1966) on theoretical geography for instance, promoted a view that cartography could be the mathematics of geography, a Cartesian space with a view from nowhere but with a view of everything (what Gregory calls the "world-as-exhibition," [Gregory 1994]). The quantitative revolution found a ready application in mapping and GIS for much the same reason as geography; it helped move the discipline away from an antiquarian approach too much concerned with trivialities and genealogies (such as classifications of historical maps). As discussed in Chapter 5, mapping's own aspirations to the status of science meant that it too needed a basis in modeling, which it found in the information theory of communication scientists such as Claude Shannon. Even as it reached for these theories (which culminated in the map communication model) however, GIS and cartography failed to see the larger picture of how these communication technologies were part of both military and corporate (or capital) spheres.

We have also seen in this book (Chapter 12) how for more than a century, but perhaps especially in the last decade or so, there has been a powerful exploration of mapping through map art, counter-mappings, and performances that has continually brought into question such things as the notion of representation and even the ontology of maps and mapping. Map artists continue to provide an incredibly rich and varied appropriation of mapping. As Wood has observed:

> Map artists . . . claim the power of the map to achieve ends other than the social reproduction of the status quo. Map artists do not reject maps. They reject the authority claimed by normative maps uniquely to portray reality as it is. (Wood 2006: n.p.)

These two directions we can think of as "resistances" (but also as I discussed in Chapter 9, "reworkings" and "resiliences") in that they set themselves as opposed to the received wisdom of the culture of mapping.

If these identified trends (securitization and resistances) in mapping and GIS mean anything it is not because this is the first time a field has experienced them ('twas ever thus no doubt), but rather because they reflect a fundamental aspect of modernity. In other words, unconstrained freedom can be dangerous – and hence the perceived need for constraints. Nor should the trends be read as definitive – Figure 1.1 is suggestive, but I recognize that it also acts to congeal one way of looking into the way of looking. The graphic, although no doubt powerful *qua* graphic, is after all only two-dimensional and static. Nor should we read a morality into it of "good" and "bad"; many technical forms of knowledge are successful, and some forms of the geoweb are markedly surveillant. Rather, this is a summary as I see it of the competing directions in which you as a user of maps and GIS are likely to find yourself being pulled.

Beyond the Cartographic Anxiety – Thinking Out Space

Gregory's conception of a cartographic anxiety over the contradictory nature of mapping and GIS was published in 1994 (Gregory 1994). It still resonates as we discussed in Chapter 1. Gregory's account of this topic was frankly pretty sprawling; it occupied some 130 pages and included postmodernism, structuration theory, Derrida, and Marxism just to name a few, and very little on mapping. But what it did do (alongside the work of Harley, Rorty, and others) was disturb the notion of mapping and GIS as "mirrors of nature" to use Rorty's phrase. A decade and a half on it's clear that critical mappings are both here to stay and hardly sufficient. Would there have been a Peters controversy, the development of GIScience ontologies, the continued appeal to representationalist science, if Gregory's critique had lodged in the heart of geography? Yet such critiques *have* opened up a spaces of alternatives, regions in which other ways of being and doing are now possible.

In an interview originally published in an architectural magazine in 1982, Michel Foucault was asked about the "political rationality" of institutions such as the Ecole nationale des ponts et chaussées. The Ecole, as Edney points out in an important study of the imperial map (Edney 2009), has been one of the main centers for training topographic engineers in France since its creation in the eighteenth century. "Those were the people," Foucault says, "who thought out space" (Foucault 1984: 244). The irony of the imperial map is that it is not something that exists in and of itself; the imperial map is the set of practices, not a sheet of paper. Imperialism is not a property of the map itself. Maps operate within a whole nexus of relations, discourses, power relationships, and material circumstances. Sometimes those circumstances are imperial and sometimes they are counter-hegemonic. The task of critical cartography and GIS is elucidate the nature of these circumstances of map practices, to challenge unexamined assumptions, and to question in what ways maps have thought out space. We need not be anxious about cartography, only anxious about being uncritical.

Note

1   I'm thinking here of the insertion of a small section on "professional ethics" in the GIS *Body of Knowledge* (DiBiase et al. 2006) or the couple of perfunctory paragraphs on "GIS and Society" inserted in a recent textbook (Slocum et al. 2009).

References

Adams, J. (2008) Sonic Youth. *The New Yorker*, August 25: 32–9.

Agamben, G. (2005) *State of Exception* (K. Attell, trans.). Chicago: The University of Chicago Press.

Agarwal, P. (2005) Ontological Considerations in Giscience. *International Journal of Geographical Information Science* 19(5): 501–36.

Agnew, J. (2002) *Making Political Geography*. London: Arnold.

Akerman, J. R. (2009) *The Imperial Map*. Chicago: University of Chicago Press.

Amoore, L. (2006) Biometric Borders: Governing Mobilities in the War on Terror. *Political Geography* 25(3): 336–51.

Armstrong, J., and Zúniga, M. M. (2006) *Crashing the Gate. Netroots, Grassroots and the Rise of People-Powered Politics*. White River Junction, VT: Chelsea Green Publishing.

Avery, S. (1989) *Up from Washington. William Pickens and the Negro Struggle for Equality, 1900–1954*. Newark: University of Delaware Press.

Bachelard, G. (1958/1969) *The Poetics of Space* (M. Jolas, trans.). Boston: Beacon Press.

Balkin, J. (2006) What Is Access to Knowledge? Retrieved August 1, 2007, from http://balkin.blogspot.com/2006/04/what-is-access-to-knowledge.html.

Bar-Zeev, A. (2008) From Keyhole to Google Earth: An Interview with Avi Bar-Zeev. *Cartographica* 43(2).

Barnes, T. J. (2006) Geographical Intelligence: American Geographers and Research and Analysis in the Office of Strategic Services 1941–1945. *Journal of Historical Geography* 32: 149–68.

Barnes, T. J. (2008) Geography's Underworld: The Military-Industrial Complex, Mathematical Modelling and the Quantitative Revolution. *GeoForum* 39: 3–16.

Barnes, T. J., and Farish, M. (2006) Between Regions: Science, Militarism, and American Geography from World War to Cold War. *Annals of the Association of American Geographers* 96(4): 807–26.

Barthes, R. (1972) *Mythologies*. New York: Hill and Wang.

BBC. (2008) Online Maps "Wiping out History." Retrieved August 30, 2008, from http://news.bbc.co.uk/1/hi/uk/7586789.stm.

Bell Labs. (2001) Claude Shannon, Father of Information Theory, Dies at 84. *Bell Labs News* 26(1).

Bentham, J. B. M. (1995) *The Panopticon Writings*. London: New York.

Bernhard, B. (2007) Mom's Milk Fuels Fight. *Orange County Register,* February 7.

Biltcliffe, P. (2005) Walter Crane and the *Imperial Federation Map Showing the Extent of the British Empire* (1886). *Imago Mundi* 57(1): 63–9.

Blomley, N. (2006) Uncritical Critical Geography? *Progress in Human Geography* 30(1): 87–94.

Board, C. (1967) Maps as Models. In R. J. Chorley and P. Haggett (eds.), *Models in Geography* (pp. 671–725). London: Methuen & Co.

Boulton, A. (2009) *Citizens as Sensors, Citizens as Censors. Or, Towards a Poststructuralist Political-Economy of Unwitting Participation, Hospitality and Censorship in Google Maps.* Paper presented at the Association of American Geographers Annual Conference, Las Vegas, NV.

Bowker, G. C., and Star, S. L. (1999) *Sorting Things Out. Classification and Its Consequences.* Cambridge, MA and London, UK: The MIT Press.

Brewer, C. A., and Suchan, T. (2001) *Mapping Census 2000. The Geography of US Diversity.* Redlands, CA: ESRI Press.

Bryan, J. (2009) Where Would We Be Without Them? Knowledge, Space and Power in Indigenous Politics. *Futures* 41: 24–32.

Buisseret, D. (1984) The Cartographic Definition of France's Eastern Boundary in the Early Seventeenth Century. *Imago Mundi* 36: 72–80.

Buisseret, D. (ed.). (1992) *Monarchs, Ministers and Maps: The Emergence of Cartography as a Tool of Government in Early Modern Europe.* Chicago: University of Chicago Press.

Buisseret, D. (2003) *The Mapmakers' Quest. Depicting New Worlds in Renaissance Europe.* Oxford: Oxford University Press.

Bunge, W. (1966) *Theoretical Geography* (2nd [rev. and enl.] edn.). Lund: Royal University of Lund Dept. of Geography Gleerup.

Burchell, G., Gordon, C., and Miller, P. (eds.). (1991) *The Foucault Effect: Studies in Governmentality: With Two Lectures by and an Interview with Michel Foucault.* Chicago: University of Chicago Press.

Butler, D. (2006) Mashups Mix Data into Global Service. *Nature* 439(7072): 6–7.

Carmi, S., Havlin, S., Kirkpatrick, S., Shavitt, Y., and Shir, E. (2007) A Model of Internet Topology Using K-Shell Decomposition. *Proceedings of the National Academy of Sciences of the United States of America* 104(27): 11150–4.

Carrubber's Mission. (1983) Carrubber's Mission Minutes: Carrubber's Christian Centre, Edinburgh, Scotland.

Casey, E. S. (2002) *Representing Place: Landscape Painting and Maps*. Minneapolis: University of Minnesota Press.

Casey, E. S. (2005) *Earth-Mapping*. Minneapolis: University of Minneapolis.

Castree, N. (2000) Professionalisation, Activism, and the University: Whither "Critical Geography"? *Environment and Planning A* 32: 955–70.

Chakraborty, J., and Bosman, M. M. (2005) Measuring the Digital Divide in the United States: Race, Income, and Personal Computer Ownership. *The Professional Geographer* 57(3): 395–410.

Chorley, R. J., and Haggett, P. (1967) *Models in Geography*. London: Methuen.

Chrisman, N. (2006) *Charting the Unknown. How Computer Mapping at Harvard Became GIS.* Redlands, CA: ESRI Press.

Christensen, K. (1982) Geography as a Human Science: A Philosophic Critique of the Positivist–Humanist Split. In P. Gould and G. Olsson (eds.), *A Search for Common Ground* (pp. 37–57). London: Pion.

Chua, H. F., Boland, J. E., and Nisbett, R. E. (2005) Cultural Variation in Eye Movements During Scene Perception. *Proceedings of the National Academy of Sciences* 102(35): 12629–33.

Clarke, K. C. (2003) *Getting Started with Geographic Information Systems* (4th edn.). Upper Saddle River, NJ and London: Pearson Education.

Cloke, P., and Johnston, R. (eds.). (2005) *Spaces of Geographical Thought*. London and Thousand Oaks, CA: Sage Publications.

Cloud, J. (2002) American Cartographic Transformations During the Cold War. *Cartography and Geographic Information Science* 29(3): 261–82.

Clute, J., and Nicholls, P. (eds.). (1995) *The Encyclopedia of Science Fiction*. New York: St. Martin's Green.

CNN. (2006) Poll: Fifth of Americans Think Calls Have Been Monitored (Vol. 2006). Washington, DC.

Colb, S. F. (2001). The New Face of Racial Profiling: How Terrorism Affects the Debate. Retrieved September 20, 2007, from http://writ.findlaw.com/colb/20011010.html.

Committee on Beyond Mapping. (2006) *Beyond Mapping. Meeting National Needs through Enhanced Geographic Information Science*. Washington, DC: The National Academies Press.

Connor, S. (2005) From 2006 Britain Will Be the First Country Where Every Journey by Every Car Will Be Monitored. *The Independent,* December 22, pp. 1, 2.

Cook, K. S. (2005) A Lifelong Curiosity About Maps. *Cartographic Perspectives* (51): 43–54.

Cosgrove, D. (ed.). (1999) *Mappings*. London: Reaktion Books.

Cosgrove, D. (2001) *Apollo's Eye: A Cartographic Genealogy of the Earth in the Western Imagination*. Baltimore, MD: Johns Hopkins University Press.

Cosgrove, D. (2005) Maps, Mapping, Modernity: Art and Cartography in the Twentieth Century. *Imago Mundi* 57(1): 35–54.

Cosgrove, D. (2006) Art and Mapping: An Introduction. *Cartographic Perspectives* (53): 4.

Cosgrove, D. E., and Della Dora, V. (2005) Mapping Global War: Los Angeles, the Pacific, and Charles Owens's Pictorial Cartography. *Annals of the Association of American Geographers* 95(2): 373–90.

Cragg, T. (n.d.) Cutting up Material. Retrieved October 9, 2009, from www.tony-cragg.com/texte/Cutting up Material.pdf.

Craig, W. J., Harris, T. M., and Weiner, D. (eds.). (2002) *Community Participation and Geographic Information Systems*. London: Taylor & Francis.

Crampton, J. W. (2003) *The Political Mapping of Cyberspace*. Chicago: University of Chicago Press.

Crampton, J. W. (2006) The Cartographic Calculation of Space: Race Mapping and the Balkans at the Paris Peace Conference of 1919. *Social and Cultural Geography* 7(5): 731–52.

Crampton, J. W. (2007a) The Biopolitical Justification for Geosurveillance. *Geographical Review* 97(3): 389–403.

Crampton, J. W. (2007b) Maps, Race and Foucault: Eugenics and Territorialization Following World War One. In J. W. Crampton and S. Elden (eds.), *Space, Knowledge and Power: Foucault and Geography* (pp. 223–44). Aldershot: Ashgate.

Crampton, J. W. (2009a) Being Ontological. *Environment and Planning D: Society and Space*: 603–8.

Crampton, J. W. (2009b) Rethinking Maps and Identity. Choropleths, Clines and Biopolitics. In M. Dodge, R. Kitchin and C. Perkins (eds.), *Rethinking Maps* (pp. 26–49). London: Routledge.

Cross, M. (2007) Copyright Fight Sinks Virtual Planning. *The Guardian,* January 4.

Cutter, S. L., Richardson, D. B., and Wilbanks, T. J. (eds.). (2003) *The Geographical Dimensions of Terrorism*. London and New York: Routledge.

Cvijić, J. (1918) The Zones of Civilization of the Balkan Peninsula. *Geographical Review* 5(6): 470–82.

Danielewski, M. Z. (2000) *House of Leaves by Zampano, with an Introduction and Notes by Johnny Truant*. New York: Pantheon Books.

de Certeau, M. (1984) *The Practice of Everyday Life*. Berkeley and Los Angeles: University of California Press.

Debord, G. (1967/1994) *Society of the Spectacle* (D. Nicolson-Smith, trans.). New York: Zone Books.

Delamarre, L. (1909) Pierre-Charles-François Dupin, *The Catholic Encyclopedia*. Robert Appleton Company.

Devlin, K. (1983) How to Put the World Back in Its Right Place. *The Guardian*, 1 December, p. 16.

DiBiase, D., DeMers, M., Johnson, A., Kemp, K., Luck, A. T., Plewe, B., and Wentz, E. (2006) *Geographic Information Science Body of Knowledge*. Washington, DC: Association of American Geographers.

DiBona, C., Cooper, D., and Stone, M. (2006) *Open Sources 2.0: The Continuing Revolution*. Sebastopol, CA: O'Reilly Media Inc.

Dobson, J. E. (2006) Geoslavery. In B. Warf (ed.), *Encyclopedia of Human Geography* (pp. 186–87). Thousand Oaks, CA: Sage Publications.

Dodds, K., and Sidaway, J. D. (2004) Halford Mackinder and the "Geographical Pivot of History": A Centennial Retrospective. *Geographical Journal* 170(4): 292–7.

Dodge, M., and Kitchin, R. (2001) *Atlas of Cyberspace*. Harlow, England: New York: Addison-Wesley.

Dodge, M., and Kitchin, R. (2007) "Outlines of a World Coming into Existence": Pervasive Computing and the Ethics of Forgetting. *Environment and Planning B: Planning and Design* 34(3): 431–45.

Dodge, M., and Perkins, C. (2008) Reclaiming the Map: British Geography and Ambivalent Cartographic Practice. *Environment and Planning A* 40(6): 1271–6.

Dominian, L. (1917) *The Frontiers of Language and Nationality in Europe*. New York: Henry Holt for the American Geographical Society.

Downs, R. M. (1994) Being and Becoming a Geographer – An Agenda for Geography Education. *Annals of the Association of American Geographers* 84(2): 175–91.

Downs, R. M. (1997) The Geographic Eye: Seeing through GIS? *Transactions in GIS* 2(2): 111–21.

Downs, R. M., and Liben, L. S. (1988) Through a Map Darkly: Understanding Maps as Representations. *Genetic Epistemologist* 16: 11–18.

Downs, R. M., and Liben, L. S. (1991) The Development of Expertise in Geography: A Cognitive-Developmental Approach to Geographic Education. *Annals of the Association of American Geographers* 81(2): 304–27.

Dreyfus, H. L. (2005) Heidegger's Ontology of Art. In H. L. Dreyfus and M. A. Wrathall (eds.), *A Companion to Heidegger* (pp. 407–19). Malden, MA and Oxford, UK: Blackwell Publishing.

Duncan, I. (1860/1869) *Pre-Adamite Man. The Story of Our Old Planet and Its Inhabitants*. 3rd edn. London: Saunders, Otley, and Co.; Edinburgh: W. P. Kennedy.

Dunn, C. E. (2007) Participatory GIS a People's GIS? *Progress in Human Geography,* 31(5): 616–37.

Dupin, P. C. F. (1827) *Forces productives et commerciales de la France* (Vol. 2). Paris: Bachelier, Libraire.

Duster, T. (2005) Race and Reification in Science. *Science* 307(5712): 1050–1.

Economist. (1989) The World Turned Upside Down. *Economist* 97.

Edney, M. H. (1992) Harley, J. B. (1932–1991) – Questioning Maps, Questioning Cartography, Questioning Cartographers. *Cartography and Geographic Information Systems* 19(3): 175–8.

Edney, M. H. (1993) Cartography Without "Progress": Reinterpreting the Nature and Historical Development of Mapmaking. *Cartographica* 30(2/3): 54–68.

Edney, M. H. (1997) *Mapping an Empire: The Geographical Construction of British India, 1765–1843*. Chicago: University of Chicago Press.

Edney, M. H. (2005a) The Origins and Development of J. B. Harley's Cartographic Theories. *Cartographica* 40(1&2): 1–143.

Edney, M. H. (2005b) Putting "Cartography" into the History of Cartography: Arthur H. Robinson, David Woodward, and the Creation of a Discipline. *Cartographic Perspectives* (51): 14–29.

Edney, M. H. (2009) The Irony of Imperial Mapping. In J. R. Akerman (ed.), *The Imperial Map* (pp. 11–45). Chicago: University of Chicago Press.

Elden, S. (2007) Governmentality, Calculation, Territory. *Environment and Planning D: Society and Space* 25(3): 562–80.

Elwood, S. (2006a) Beyond Cooptation or Resistance: Urban Spatial Politics, Community Organizations, and GIS-Based Spatial Narratives. *Annals of the Association of American Geographers* 96(2): 323–41.

Elwood, S. (2006b) Critical Issues in Participatory GIS: Deconstructions, Reconstructions, and New Research Directions. *Transactions in GIS* 10(5): 693–708.

Elwood, S. (2008) Volunteered Geographic Information: Future Research Directions Motivated by Critical, Participatory, and Feminist GIS. *GeoJournal*: 1–11.

Eribon, D. (2004) *Insult and the Making of the Gay Self*. Durham, NC: Duke University Press.

Erle, S., Gibson, R., and Walsh, J. (2005) *Mapping Hacks*. Sebastopol, CA: O'Reilly and Associates.

Fairhurst, R. (2005) Next-Generation Webmapping. *Bulletin of the Society of University Cartographers* 39(1–2): 57–61.

Foucault, M. (1977) *Discipline and Punish: The Birth of the Prison* (1st American edn.). New York: Pantheon Books.

Foucault, M. (1978) *The History of Sexuality* (1st American edn.). New York: Pantheon Books.

Foucault, M. (1983) The Subject and Power. In H. L. Dreyfus and P. Rabinow (eds.), *Michel Foucault: Beyond Structuralism and Hermeneutics* (2nd edn., pp. 208–26). Chicago: University of Chicago Press.

Foucault, M. (1984) Space, Knowledge, and Power. In P. Rabinow (ed.), *The Foucault Reader* (pp. 239–56). New York: Pantheon.

Foucault, M. (1985) *The Use of Pleasure. The History of Sexuality*, Vol. 2. New York: Vintage.

Foucault, M. (1997a) The Ethics of the Concern for Self as a Practice of Freedom. In P. Rabinow (ed.), *Ethics, Subjectivity and Truth. Essential Works of Foucault 1954–1984*, Vol. I (pp. 281–301). New York: The New Press.

Foucault, M. (1997b) *The Politics of Truth*. New York: Semiotext(e) (Distributed by the MIT Press).

Foucault, M. (2000a) About the Concept of The "Dangerous Individual" in Nineteenth-Century Legal Psychiatry. In J. D. Faubion (ed.), *Essential Foucault: Power* (pp. 176–200). New York: New Press.

Foucault, M. (2000b) "Omnes Et Singulatim": Toward a Critique of Political Reason. In J. Faubion (ed.), *Power. The Essential Works of Michel Foucault 1954–1984*, Vol. 3 (pp. 298–325). New York: New Press.

Foucault, M. (2000c) So Is It Important to Think? In J. D. Faubion (ed.), *Power. The Essential Works of Michel Foucault 1954–1984*, Vol. 3 (pp. 454–8). New York: The New Press.

Foucault, M. (2000d) Truth and Juridical Forms. In J. Faubion (ed.), *Power. The Essential Works of Michel Foucault 1954–1984*, Vol. 3 (pp. 1–89). New York: New Press.

Foucault, M. (2003a) *Abnormal: Lectures at the Collège de France (1974–1975)*. New York: Picador.

Foucault, M. (2003b) *Society Must Be Defended: Lectures at the Collège de France, 1975–76*. New York: Picador.

Foucault, M. (2007) *Security, Territory, and Population. Lectures at the Collège de France* (G. Burchell, trans.). Basingstoke and New York: Palgrave Macmillan.

Foucault, M. (2008) *The Birth of Biopolitics. Lectures at the Collège de France 1978–1979*. Basingstoke and New York: Palgrave Macmillan.

Foucault, M., Martin, L. H., Gutman, H., and Hutton, P. H. (1988) *Technologies of the Self: A Seminar with Michel Foucault*. Amherst: University of Massachusetts Press.

Foucault, M., and Pearson, J. (2001) *Fearless Speech*. Los Angeles, CA: Semiotext(e) (Distributed by the MIT Press).

Friedman, J. B. (1981) *The Monstrous Races in Medieval Art and Thought*. Cambridge, MA: Harvard University Press.

Friendly, M. (2002) Visions and Re-Visions of Charles Joseph Minard. *Journal of Educational and Behavioral Statistics* 27(1): 31–51.

FRUS. (1942–7) *Papers Relating to the Foreign Relations of the United States. The Paris Peace Conference, 1919*. Washington, DC: US Govt. Print. Off.

Gall, J. (1856) On Improved Monographic Projections of the World. *Report of the Twenty-Fifth Meeting of the British Association for the Advancement of Science*: 148.

Gall, J. (1860[1858]) *The Stars and Angels*. Philadelphia: William S. and Alfred Martien.

Gall, J. (1871) On a New Projection for a Map of the World. *Royal Geographical Society Proceedings* 15(July 12): 159.

Gall, J. (1880) *Primeval Man Unveiled: Or, the Anthropology of the Bible*. 2nd edn. London: Hamilton, Adams and Co.

Gall, J. (1885) Use of Cylindrical Projections for Geographical, Astronomical, and Scientific Purposes. *Scottish Geographical Magazine* 1: 119–23.

Gantz, J. F. (2008) *The Diverse and Exploding Digital Universe*. Framingham, MA: IDC.

Gigerenzer, G. (2004) Dread Risk, September 11, and Fatal Traffic Accidents. *Psychological Science* 15(4): 286–7.

Gigerenzer, G. (2006) Out of the Frying Pan into the Fire: Behavioral Reactions to Terrorist Attacks. *Risk Analysis* 26(2): 347–51.

Giles, J. (2005) Internet Encyclopedias Go Head to Head. *Nature* 438 (15 December): 900–1.

Gleick, J. (2001) Bit Player. *The New York Times Magazine*, December 30, p. 48.

Goodchild, M. (1992) Geographical Information Science. *International Journal of Geographical Information Systems* 6(1): 31–45.

Goodchild, M. (2007) Citizens as Censors: The World of Volunteered Geography. *GeoJournal* 69: 211–21.

Goodman, A. H. (2006) Two Questions About Race. Retrieved August 17, 2008, from http://raceandgenomics.ssrc.org/Goodman/.

Goodman, A. H., Heath, D., and Lindee, M. S. (eds.). (2003) *Genetic Nature/Culture. Anthropology and Science Beyond the Two-Culture Divide*. Berkeley, CA: University of California Press.

Gordon, C. (1991) Governmental Rationality: An Introduction. In G. Burchell, C. Gordon and P. Miller (eds.), *The Foucault Effect: Studies in Governmentality* (pp. 1–51). Chicago: University of Chicago Press.

Gore, A. (1998) The Digital Earth: Understanding Our Planet in the 21st Century. Retrieved September 15, 2007, from http://www.isde5.org/al_gore_speech.htm.

Gore, A. (2007) *The Assault on Reason*. New York: Penguin Press.

Grant, M. (1916) The Passing of the Great Race. *Geographical Review* 2(5): 354–60.

Grant, M. (1917) Introduction. In L. Dominian (ed.), *The Frontiers of Language and Nationality in Europe* (pp. xii–xviii). New York: Henry Holt and Company.

Grant, M. (1932) *The Passing of the Great Race or the Racial Basis of European History* (4th edn.). New York: Scribner's Sons.

Graves, J. L. (2001) *The Emperor's New Clothes. Biological Theories of Race at the Millennium*. New Brunswick, NJ: Rutgers University Press.

Greenhood, D. (1964) *Mapping*. Chicago: University of Chicago Press.

Greenwald, G. (2008a) Federal Government Involved in Raids on Protesters. Retrieved August 31, 2008, from http://www.salon.com/opinion/greenwald/2008/08/31/raids/.

Greenwald, G. (2008b) Massive Police Raids on Suspected Protestors in Minneapolis. Retrieved August 30, 2008, from http://www.salon.com/opinion/greenwald/2008/08/30/police_raids/.

Gregory, D. (1994) *Geographical Imaginations*. Cambridge, MA and Oxford, UK: Blackwell.

Gregory, D. (2004) *The Colonial Present. Afghanistan, Palestine, Iraq*. Malden, MA: Blackwell Publishing.

Gregory, D., and Pred, A. (2007) *Violent Geographies*. New York and London: Routledge.

Hacking, I. (1975) *The Emergence of Probability: A Philosophical Study of Early Ideas About Probability, Induction and Statistical Inference*. London and New York: Cambridge University Press.

Hacking, I. (1982) Biopower and the Avalanche of Printed Numbers. *Humanities in Society* 5: 279–95.

Hacking, I. (1990) *The Taming of Chance*. Cambridge, UK and New York: Cambridge University Press.

Hacking, I. (2002) *Historical Ontology*. Cambridge, MA: Harvard University Press.

Hall, M. (2007) On the Mark: Will Democracy Vote the Experts Off the GIS Island?, *Computerworld* (Vol. 2007). Framington, MA.

Hannah, M. (2000) *Governmentality and the Mastery of Territory in Nineteenth-Century America*. Cambridge: Cambridge University Press.

Hannah, M. (2006) Torture and the Ticking Bomb: The War on Terrorism as a Geographical Imagination of Power/Knowledge. *Annals of the Association of American Geographers* 96(3): 622–40.

Hannah, M. (2009) Calculable Territory and the West German Census Boycott Movements of the 1980s. *Political Geography* 28(1): 66–75.

Harley, J. B. (1964) *The Historian's Guide to Ordnance Survey Maps: Reprinted from the Amateur Historian with Additional Material*. London: Published for the Standing Conference for Local History by the National Council of Social Service.

Harley, J. B. (1969–71) Bibliographical Notes. In *The First Edition of One-Inch Ordnance Survey of England and Wales*. Newton Abbott, Devon: David & Charles.

Harley, J. B. (1987) The Map as Biography: Thoughts on Ordnance Survey Map, Six-Inch Sheet Devonshire Cix, Se, Newton Abbot. *The Map Collector* (41): 18–20.

Harley, J. B. (1988a) Maps, Knowledge, and Power. In D. Cosgrove and S. Daniels (eds.), *The Iconography of Landscape: Essays on the Symbolic Representation, Design and Use of Past Environments* (pp. 277–312). Cambridge: Cambridge University Press.

Harley, J. B. (1988b) Silences and Secrecy: The Hidden Agenda of Cartography in Early Modern Europe. *Imago Mundi* 40: 57–76.

Harley, J. B. (1989a) Deconstructing the Map. *Cartographica* 26(2): 1–20.

Harley, J. B. (1989b) "The Myth of the Great Divide": Art, Science, and Text in the History of Cartography. *13th International Conference on the History of Cartography*. Amsterdam.

Harley, J. B. (1990a) Cartography, Ethics and Social Theory. *Cartographica* 27(2): 1–23.

Harley, J. B. (1990b) *Maps and the Columbian Encounter: An Interpretive Guide to the Travelling Exhibition*. University of Wisconsin-Milwaukee, Golda Meir Library.

Harley, J. B. (1991) Can There Be a Cartographic Ethics? *Cartographic Perspectives* 10: 9–16.

Harley, J. B. (1992a) Deconstructing the Map. In T. J. Barnes and J. S. Duncan (eds.), *Writing Worlds: Discourse, Text and Metaphor in the Representation of Landscape* (pp. 231–47). London: Routledge.

Harley, J. B. (1992b) Rereading the Maps of the Columbian Encounter. *Annals of the Association of American Geographers* 82(3): 522–42.

Harley, J. B. (2001) *The New Nature of Maps: Essays in the History of Cartography*. Baltimore, MD: Johns Hopkins University Press.

Harley, J. B., and Woodward, D. (eds.). (1987) *Cartography in Prehistoric, Ancient, and Medieval Europe and the Mediterranean*. Chicago: University of Chicago Press.

Harley, J. B., and Zandvliet, K. (1992) Art, Science, and Power in Sixteenth-Century Dutch Cartography. *Cartographica* 29(2): 10–19.

Harmon, K. (2004) *You Are Here. Personal Geographies and Other Maps of the Imagination*. New York: Princeton Architectural Press.

Harris, L. M., and Hazen, H. D. (2006) Power of Maps: (Counter) Mapping for Conservation. *ACME* 4(1): 99–130.

Hartshorne, R. (1939) *The Nature of Geography: A Critical Survey of Current Thought in the Light of the Past*. Lancaster, PA: Association of American Geographers.

Harvey, D. (1990) *The Condition of Postmodernity: An Enquiry into the Origins of Cultural Change*. Oxford, UK and Malden, MA: Blackwell Publishers.

Harvey, D. (2001) Cartographic Identities: Geographical Knowledges under Globalization. In D. Harvey (ed.), *Spaces of Capital: Towards a Critical Geography* (pp. 208–33). New York: Routledge.

Harvey, D. (2003) *Paris, Capital of Modernity*. New York: Routledge.

Hayes, P. G. (1995) Plotting Our Past. A UW Prof Tackles the History of Maps. *The Milwaukee Journal*.

Heidegger, M. (1962) *Being and Time*. New York: Harper.

Heidegger, M. (1977) *The Question Concerning Technology, and Other Essays* (W. Lovitt, trans.). New York: Harper and Row.

Heidegger, M. (1993) The Origin of the Work of Art. In D. F. Krell (ed.), *Martin Heidegger Basic Writings*. Rev. edn. (pp. 143–212). New York: HarperCollins.

Helft, M. (2007) With Simple Tools on Web, Amateurs Reshape Mapmaking. *The New York Times*, July 27, p. A1.

Holdich, T. H. (1916) Geographical Problems in Boundary Making. *The Geographical Journal* 47(6): 421–36.

Holloway, S. L., Rice, S. P., and Valentine, G. (eds.). (2003) *Key Concepts in Geography*. London: Sage Publications.

Huxley, M. (2006) Spatial Rationalities: Order, Environment, Evolution and Government. *Social and Cultural Geography* 7(5): 771–87.

Isikoff, M. (2003) The FBI Says, Count the Mosques. *Newsweek*, February 3, p. 6.

Jacob, C. (2006) *The Sovereign Map: Theoretical Approaches in Cartography through History* (T. Conley, trans.). Chicago: University of Chicago Press.

Jefferson, M. (1909) The Anthropography of Some Great Cities: A Study in Distribution of Population. *Bulletin of the American Geographical Society* 41(9): 537–66.

Jessop, B. (2007) From Micro-Powers to Governmentality: Foucault's Work on Statehood, State Formation, Statecraft and State Power. *Political Geography* 26(1): 34–40.

Johnson, D. W. (1919) A Geographer at the Front and at the Peace Conference. *Natural History* XIX(6): 511–17.

Johnson, F. C., and Klare, G. R. (1961) General Models of Communication Research, a Survey of the Development of the Decade. *Journal of Communication* 11(1): 13–26, 45.

Johnson, S. (2006) *The Ghost Map*. New York: Riverhead Books.

Johnston, N. (1994) *Eastern State Penitentiary: Crucible of Good Intentions*. Philadelphia, PA: Philadelphia Museum of Art.

Johnston, R. (2001) Out of the "Moribund Backwater": Territory and Territoriality in Political Geography. *Political Geography* 20(6): 677–93.

Jones, M., Jones, R., and Woods, M. (2004) *An Introduction to Political Geography. Space, Place and Politics*. London and New York: Routledge.

Jordan, T. G. (1988) The Intellectual Core. *AAG Newsletter* 23(5): 1.

Kahn, J. (2007) Race in a Bottle. *Scientific American* 297(2): 40–5.

kanarinka. (2006) Art-Machines, Body-Ovens and Map-Recipes: Entries for a Psychogeographic Dictionary. *Cartographic Perspectives* (53): 24–40.

kanarinka. (2009) Art and Cartography. In R. Kitchin and N. Thrift (eds.), *The International Encyclopedia of Human Geography* (pp. 190–206). Oxford: Elsevier.

Kant, I. (1781; 2nd edn. 1787) *Critique of Pure Reason*.

Kant, I. (2001/1784) What Is Enlightenment? In A. W. Wood (ed.), *Basic Writings of Kant* (pp. 133–41). New York: The Modern Library.

Katz, B. M. (1989) *Foreign Intelligence. Research and Analysis in the Office of Strategic Services 1942–1945*. Cambridge, MA: Harvard University Press.

Katz, C. (2004) *Growing up Global. Economic Restructuring and Children's Everyday Lives*. Minneapolis, MN: University of Minnesota Press.

Keen, A. (2007) *The Cult of the Amateur. How Today's Internet Is Killing Our Culture*. New York: Doubleday/Currency.

King, M. (2006) Bottled Water Forces Flight to Land. Retrieved February 14, 2007, from http://www.11alive.com/news/news_article.aspx?storyid=83684.

Kitchin, R., and Dodge, M. (2007) Rethinking Maps. *Progress in Human Geography* 31(3): 331–44.

Klinkenberg, B. (2007) Geospatial Technologies and the Geographies of Hope and Fear. *Annals of the Association of American Geographers* 97(2): 350–60.

Koláčný, A. (1969) Cartographic Information – a Fundamental Concept and Term in Modern Cartography. *Cartographic Journal* 6: 47–9.

Konvitz, J. W. (1987) *Cartography in France 1660–1848: Science, Engineering, and Statecraft.* Chicago: University of Chicago Press.

Krogt, P. v. d. (2006) "Kartografie" or "Cartografie"? *Caart-Thresoor* 25(1): 11–12.

Krygier, J. (1996) Geography and Cartographic Design. In C. Wood and C. P. Keller (eds.), *Cartographic Design: Theoretical and Practical Perspectives* (pp. 19–33). New York: Wiley.

Kwan, M. P. (2002a) Feminist Visualization: Re-Envisioning GIS as a Method in Feminist Geographic Research. *Annals of the Association of American Geographers* 92(4): 645–61.

Kwan, M. P. (2002b) Is GIS for Women? Reflections on the Critical Discourse of the 1990s. *Gender, Place and Culture* 9(3): 271–9.

Kwan, M. P. (2007) Affecting Geospatial Technologies: Toward a Feminist Politics of Emotion. *Professional Geographer* 59(1): 22–34.

Kwan, M. P., and Ding, G. (2008) Geo-Narrative: Extending Geographic Information Systems for Narrative Analysis in Qualitative and Mixed-Method Research. *The Professional Geographer* 60(4): 443–65.

Kwan, M. P., and Schuurman, N. (2004) Introduction: Issues of Privacy Protection and Analysis of Public Health Data. *Cartographica* 39(2): 1–4.

Lagos, M. (2006) Diverted Flight Arrives in S.F. *San Francisco Chronicle*, September 11.

Latour, B. (2004) Why Has Critique Run out of Steam? From Matters of Fact to Matters of Concern. *Critical Inquiry* 30: 225–48.

Leszczynski, A. (2009a) Poststructuralism and GIS: Is There a "Disconnect". *Environment and Planning D: Society and Space* 27(4): 581–602.

Leszczynski, A. (2009b) Quantitative Limits to Qualitative Discussions: GIS, Its Critics, and the Philosophical Divide. *The Professional Geographer* 61(3): 350–65.

Lewis, G. M. (1992) Milwaukee and the American Encounter. In *A Celebration of the Life and Work of J. B. Harley 1932–1991 [17 March 1992]* (pp. 16–19). London: Royal Geographical Society.

Lewis, P. (1992) Introducing a Cartographic Masterpiece: A Review of the U.S. Geological Survey's Digital Terrain Map of the United States, by Gail Thelin and Richard Pike. *Annals of the Association of American Geographers* 82(2): 289–99.

Lewontin, R. C. (1972) The Apportionment of Human Diversity. In M. K. Hecht and W. S. Steere (eds.), *Evolutionary Biology*, Vol. 6 (pp. 381–98). New York: Plenum.

Li, W., Yang, C., and Raskin, R. (2008) A Semantic Enhanced Search for Spatial Web Portals. Paper presented at the AAAI Spring Symposium Technical Report, SS-08-05: 47–50.

Liben, L. S., and Downs, R. M. (1989) Understanding Maps as Symbols: The Development of Map Concepts in Children. In H. Reese (ed.), *Advances in Child Development and Behavior*, Vol. 22 (pp. 145–201). New York: Academic Press.

Livingstone, D. N. (1992a) *The Geographical Tradition.* Oxford: Blackwell.

Livingstone, D. N. (1992b) The Preadamite Theory and the Marriage of Science and Religion. *Transactions of the American Philosophical Society* 82(3): v–x, 1–81.

Livingstone, D. N. (2003) *Putting Science in Its Place: Geographies of Scientific Knowledge.* Chicago: University of Chicago Press.

Livingstone, D. N. (2008) *Adam's Ancestors.* Baltimore: The Johns Hopkins University Press.

Lorimer, H. (2008) Cultural Geography: Non-Representational Conditions and Concerns. *Progress in Human Geography* 32(4): 551–9.

Lyell, C. (1863) *The Geological Evidences of the Antiquity of Man with Remarks on Theories of the Origin of Species by Variation.* London: John Murray.

Lyman, P., and Varian, H. R. (2003) How Much Information. Retrieved March 14, 2008, from http://www2.sims.berkeley.edu/research/projects/how-much-info-2003/.

Lyon, D. (1994) *The Electronic Eye. The Rise of the Surveillance Society*. Minneapolis, MN: University of Minnesota Press.

Lyon, D. (2003) Surveillance Technology and Surveillance Society. In T. J. Misa, P. Brey and A. Feenberg (eds.), *Modernity and Technology* (pp. 161–83). Cambridge, MA: MIT Press.

MacEachren, A. M. (1998) Cartography, GIS and the World Wide Web. *Progress in Human Geography* 22(4): 575–85.

MacEachren, A. M., and Brewer, I. (2004) Developing a Conceptual Framework for Visually-Enabled Geocollaboration. *International Journal of Geographical Information Science* 18(1): 1–34.

MacEachren, A. M., Cai, G., Sharma, R., Rauschert, I., Brewer, I., Bolelli, L., Shaparenko, B., Fuhrmann, S., and Wang, H. (2005) Enabling Collaborative Geoinformation Access and Decision-Making through a Natural, Multimodal Interface. *International Journal of Geographical Information Science* 19(3): 293–317.

MacEachren, A. M., Pike, W., Yu, C., Brewer, I., Gahegan, M., Weaver, S. D., and Yarnal, B. (2006) Building a Geocollaboratory: Supporting Human-Environment Regional Observatory (Hero) Collaborative Science Activities. *Computers, Environment and Urban Systems* 30(2): 201–25.

Mackinder, H. J. (1904) The Geographical Pivot of History. *The Geographical Journal* 23(4): 421–37.

MAPPS. (2007) QBS Litigation Update. Retrieved June 15, 2007, from http://www.mapps.org/QBSlawsuit.asp.

Mark, D., and Turk, A. G. (2003) Landscape Categories in Yindjibarndi: Ontology, Environment, and Language. *Lecture Notes in Computer Science (including subseries Lecture Notes in Artificial Intelligence and Lecture Notes in Bioinformatics)* 2825: 28–45.

Mark, D., Turk, A. G., and Stea, D. (2007) Progress on Yindjibarndi Ethnophysiography. *Lecture Notes in Computer Science (including subseries Lecture Notes in Artificial Intelligence and Lecture Notes in Bioinformatics)*, 4736: 1–19.

Marks, J. (1995) *Human Biodiversity*. Hawthorne, NY: Aldine de Gruyter.

Marks, J. (2006) The Realities of Races. Retrieved March 15, 2008, from http://raceandgenomics.ssrc.org/Marks/.

Martin, G. J. (1968) *Mark Jefferson, Geographer*. Ypsilanti, MI: Eastern Michigan University Press.

Martin, L. (1946/2005) Arthur Robinson and the OSS. *Cartographic Perspectives* (51): 67.

Massey, R. (2007) Fifty Per Cent of Drivers Cannot Read a Map. *Daily Mail*, 6 August.

McAuliffe, B., and Simons, A. (2008) Police Raids Enrage Activists, Alarm Others. *Minneapolis Star Tribune*, August 31, 2008.

Merton, R. K. (1968) The Matthew Effect in Science. *Science*, 159(3819): 56–63.

Mindell, D., Segal, J., and Gerovitch, S. (2003) From Communications Engineering to Communications Science. In M. Walker (ed.), *Science and Ideology. A Comparative History* (pp. 66–96). London and New York: Routledge.

Miller, C. C. (2006) A Beast in the Field: The Google Maps Mashup as GIS/2. *Cartographica* 41(3): 187–99.

Misa, T. J., Brey, P., and Feenberg, A. (eds.). (2003) *Modernity and Technology*. Cambridge, MA: The MIT Press.

Mitchell, D. (2000) *Cultural Geography: A Critical Introduction*. Malden, MA: Blackwell Publishing.

Monmonier, M. (1985) *Technological Transition in Cartography*. Madison, WI: University of Wisconsin Press.

Monmonier, M. (1989) *Maps with the News: The Development of American Journalistic Cartography*. Chicago: University of Chicago Press.

Monmonier, M. (1991) *How to Lie with Maps*. Chicago: University of Chicago Press.

Monmonier, M. (1995) *Drawing the Line: Tales of Maps and Cartocontroversy* (1st edn.). New York: H. Holt.

Monmonier, M. (1997) *Cartographies of Danger: Mapping Hazards in America*. Chicago: University of Chicago Press.

Monmonier, M. (2001) *Bushmanders and Bullwinkles: How Politicians Manipulate Electronic Maps and Census Data to Win Elections*. Chicago: University of Chicago Press.

Monmonier, M. (2002a) Maps, Politics, and History. An Interview with Mark Monmonier Conducted by Jeremy W. Crampton. *Environment and Planning D-Society and Space* 20(6): 637–46.

Monmonier, M. (2002b) *Spying with Maps: Surveillance Technologies and the Future of Privacy*. Chicago: University of Chicago Press.

Montello, D. R. (2002) Cognitive Map-Design Research in the Twentieth Century: Theoretical and Empirical Approaches. *Cartography and Geographic Information Science* 29(3): 283–304.

Morris, J. A. (1973) Dr Peters' Brave New World. *The Guardian*, June 5, p. 15.

Morris, N. (2007) Fewer Than One in 20 Held Under Anti-Terror Laws Is Charged. *The Independent*, March 6, p. 10.

Newsweek. (2007) Newsweek Poll Conducted by Princeton Survey Research Associates. Retrieved August 31, 2008, from http://pollingreport.com/terror.htm.

Nietschmann, B. (1995) Defending the Miskito Reefs with Maps and GPS: Mapping with Sail, Scuba and Satellite. *Cultural Survival Quarterly* 18: 34–7.

NitroMed Inc. (2005) Bidil Named to American Heart Association's 2004 "Top 10 Advances" List; Only Cariovascular Drug Recognized by Aha for Dramatically Improving Survivial in African American Hearth Failure Patients. *PR Newswire US*, 11 January.

Nobles, M. (2000) *Shades of Citizenship: Race and the Censusin Modern Politics*. Stanford, CA: Stanford University Press.

Noyes, J. K. (1994) The Natives in Their Places: "Ethnographic Cartography" and the Representation of Autonomous Spaces in Ovamboland, German South West Africa. *History and Anthropology* 8(1–4): 237–64.

O' Tuathail, G. (1996) *Critical Geopolitics: The Politics of Writing Global Space*. Minneapolis: University of Minnesota Press.

ODT Inc. (Writer) (2008) Arno Peters: Radical Map, Remarkable Man [DVD]. B. Abramms (Producer). USA: ODT Maps.

Openshaw, S. (1991) A View on the GIS Crisis in Geography, or Using GIS to Put Humpty Dumpty Back Together Again. *Environment and Planning A* 23: 621–8.

Openshaw, S. (1992) Further Thoughts on Geography and GIS – a Reply. *Environment and Planning A* 24(4): 463–6.

Openshaw, S. (1997) The Truth About Ground Truth. *Transactions in GIS* 2(1): 7–24.

Ormeling, F. (1992) The Influence of Brian Harley on Modern Cartography. *Caert-Tresoor* 11(1): 2–6.

Orwell, G. (2003) *Nineteen Eighty-Four*. New York: Plume.

Oxford English Dictionary. (1989) *Oxford English Dictionary* (2nd edn.). Oxford: Oxford University Press.

Paglen, T. (2007) Unmarked Planes and Hidden Geographies. Retrieved March 15, 2007, from http://vectors.usc.edu/index.php?page=7&projectId=59.

Paglen, T., and Thompson, A. C. (2006) *Torture Taxi. On the Trail of the CIA's Rendition Flights.* Hoboken, NJ: Melville House Publishing.

Painter, J. (2006) Cartophilias and Cartoneuroses. *Area* 38(3): 345–7.

Painter, J. (2008) Cartographic Anxiety and the Search for Regionality. *Environment and Planning A* 40: 342–61.

Pavlovskaya, M. (2006) Theorizing with GIS: A Tool for Critical Geographies? *Environment and Planning A* 38(11): 2003–20.

Pavlovskaya, M. (2009) Critical GIS and Its Positionality. *Cartographica* 44(1): 8–10.

Pearce, M. (2006) Narrative Structures for Cartographic Design. Paper presented at the Association of American Geographers Annual Conference, Chicago.

Perec, G. (1974/1997) *Species of Spaces and Other Pieces* (J. Sturrock, trans.). London: Penguin Books.

Perkins, C. (2003) Cartography: Mapping Theory. *Progress in Human Geography* 27(3): 341–51.

Perlmutter, D. D. (2006) Are Bloggers "The People"? Retrieved January 27, 2007, from http://policybyblog.squarespace.com/are-bloggers-the-people/.

Perlmutter, D. D. (2008) *Blogwars. The New Political Battleground.* Oxford: Oxford University Press.

Peters, A. (1974) The Europe-Centered Character of Our Geographic View of the World and Its Correction. Retrieved February 11, 2006, from http://www.heliheyn.de/Maps/Lect02_E.html.

Peters, A. (1983) *The New Cartography.* New York: Friendship Press.

Petto, C. M. (2005) From l'état, c'est moi to l'état, c'est l'état: Mapping in Early Modern France. *Cartographica* 40(3): 53–78.

Pickens, W. (1991/1923) *Bursting Bonds* (enlarged edn.). Bloomington, IN: Indiana University Press.

Pickles, J. (1991) Geography, GIS, and the Surveillant Society. *Papers and Proceedings of Applied Geography Conferences* 14: 80–91.

Pickles, J. (1995) *Ground Truth.* New York: Guilford.

Pickles, J. (1999) Arguments, Debates, and Dialogues: The GIS–Social Theory Debate and the Concern for Alternatives. In P. A. Longley, M. F. Goodchild, D. J. Maguire, and D. W. Rhind (eds.), *Geographical Information Systems*, Vol. 1 (pp. 49–60). New York: John Wiley.

Pickles, J. (2004) *A History of Spaces. Cartographic Reason, Mapping and the Geo-Coded World.* London: Routledge.

Pickles, J. (2006) On the Social Lives of Maps and the Politics of Diagrams: A Story of Power, Seduction, and Disappearance. *Area* 38(3): 347–50.

Pinder, D. (1996) Subverting Cartography: The Situationists and Maps of the City. *Environment and Planning A* 28(3): 405–27.

Pinder, D. (2003) Mapping Worlds. Cartography and the Politics of Representation. In A. Blunt, P. Gruffudd, J. May, M. Ogborn, and D. Pinder (eds.), *Cultural Geography in Practice* (pp. 172–87). London: Arnold.

Pinder, D. (2005) *Visions of the City. Utopianism, Power and Politics in Twentieth-Century Urbanism.* Edinburgh: University of Edinburgh Press.

Pliny the Elder. (1938–63) *Natural History* (H. Rackham, trans.). Cambridge, MA: Harvard University Press.

Polt, R. F. H. (1999) *Heidegger: An Introduction.* Ithaca, NY: Cornell University Press.

Priest, C. (1978/1999) The Watched. In *The Dream Archipelago* (pp. 186–264). London: Earthlight Books.

Rainie, L., and Horrigan, J. (2007) Election 2006 Online. Retrieved February 20, 2008, from http://www.pewinternet.org/pdfs/PIP_Politics_2006.pdf.

Raisz, E. (1938) *General Cartography.* New York: McGraw-Hill.

Raskin, R. (2005) Knowledge Representation in the Semantic Web for Earth and Environmental Terminology (Sweet). *Computers and Geosciences* 31(9): 1119–25.

Ratajski, L. (1974) Commission V of the ICA: The Tasks It Faces. *International Yearbook of Cartography* 14: 140–4.

Ratliff, E. (2007) The Whole Earth, Catalogued. How Google Maps Is Changing the Way We See the World. *Wired* 15(7): 154–9.

Raymond, E. (2001) *The Cathedral and the Bazaar: Musings on Linux and Open Source by an Accidental Revolutionary* (rev. edn.). Cambridge, MA: O'Reilly Media Inc.

Read, B. (2007) Middlebury College History Department Limits Students' Use of Wikipedia. *The Chronicle of Higher Education*, February 16.

Robinson, A. H. (1952) *The Look of Maps: An Examination of Cartographic Design.* Madison: University of Wisconsin Press.

Robinson, A. H. (1953) *Elements of Cartography.* New York: John Wiley and Sons.

Robinson, A. H. (1967) The Thematic Maps of Charles Joseph Minard. *Imago Mundi* 21: 95–108.

Robinson, A. H. (1979) Geography and Cartography Then and Now. *Annals of the Association of American Geographers* 69(1): 97–102.

Robinson, A. H. (1982) *Early Thematic Mapping in the History of Cartography.* Chicago: University of Chicago Press.

Robinson, A. H. (1985) Arno Peters and His New Cartography. *The American Cartographer* 12: 103–11.

Robinson, A. H. (1991) The Development of Cartography at the University of Wisconsin-Madison. *Cartography and Geographic Information Systems* 18(3): 156–7.

Robinson, A. H. (1997) The President's Globe. *Imago Mundi* 49: 143–52.

Robinson, A. H., Morrison, J. L., and Muehrcke, P. C. (1977) Cartography 1950–2000. *Transactions of the Institute of British Geographers* NS 2(1): 3–18.

Robinson, A. H., and Petchenik, B. B. (1976) *The Nature of Maps: Essays Toward Understanding Maps and Mapping.* Chicago: University of Chicago Press.

Robinson, A. H., and Wallis, H. M. (1967) Humboldt's Map of Isothermal Lines: A Milestone in Thematic Cartography. *The Cartographic Journal* 4: 119–23.

Rorty, R. (1979) *Philosophy and the Mirror of Nature.* Princeton: Princeton University Press.

Rose, G. (2001) *Visual Methodologies: An Introduction to the Interpretation of Visual Materials.* London: Sage.

Roush, W. (2005) Killer Maps. *Technology Review* 108(10): 54–60.

Royal Geographical Society. (1992) *A Celebration of the Life and Work of J. B. Harley, 1932–1991.* RGS: London.

Rundstrom, R. A. (1995) GIS, Indigenous Peoples, and Epistemological Diversity. *Cartography and Geographic Information Systems* 22: 45–57.

Sacks, O. W. (1985) *The Man Who Mistook His Wife for a Hat and Other Clinical Tales.* New York: Summit Books.

Said, E. W. (2000) *Reflections on Exile and Other Essays.* Cambridge, MA: Harvard University Press.

Sankar, P., and Kahn, J. (2005) Bidil: Race Medicine or Race Marketing? *Health Affairs*, October 11: 455–63.

Scharl, A., and Tochtermann, K. (2007) *The Geospatial Web. How Geobrowsers, Social Software and the Web 2.0 Are Shaping the Network Society.* London: Springer.

Schuurman, N. (1999a) Critical GIS: Theorizing an Emerging Science. *Cartographica* 36(4): 1–107.

Schuurman, N. (1999b) Speaking with the Enemy? A Conversation with Michael Goodchild. *Environment and Planning D-Society and Space* 17(1): 1–2.

Schuurman, N. (2000) Trouble in the Heartland: GIS and Its Critics in the 1990s. *Progress in Human Geography* 24(4): 569–90.

Schuurman, N. (2002) Care of the Subject: Feminism and Critiques of GIS. *Gender, Place and Culture* 9(3): 291–9.

Schuurman, N. (2004) *GIS: A Short Introduction.* Malden, MA: Blackwell Publishers.

Schuurman, N., and Kwan, M. P. (2004) Guest Editorial: Taking a Walk on the Social Side of GIS. *Cartographica* 39(1): 1–3.

Science Daily. (2006) New Technology Helping Foster the "Democratization of Cartography." Retrieved March 24, 2007, from http://www.sciencedaily.com/releases/2006/09/060920192549.htm.

Scott, J. C. (1998) *Seeing Like a State: How Certain Schemes to Improve the Human Condition Have Failed.* New Haven: Yale University Press.

Shadbolt, N., and Berners-Lee, T. (2008) Web Science Emerges. *Scientific American* 299(4): 76–81.

Shannon, C. (1948) A Mathematical Theory of Communication. *The Bell System Technical Journal* 27: 379–423, 623–56.

Shaw, M., and Miles, I. (1979) The Social Roots of Statistical Knowledge. In J. Irvine, I. Miles and J. Evans (eds.), *Demystifying Social Statistics.* London: Pluto Press.

Sheppard, E. (1995) GIS and Society: Towards a Research Agenda. *Cartography and Geographic Information Systems* 22(1): 5–16.

Sheppard, E. (2005) Knowledge Production through Critical GIS: Genealogy and Prospects. *Cartographica* 40(4): 5–21.

Sheppard, E. (2009) Branding GIS: What's "Critical"? *Cartographica* 44(1): 13–14.

Siegel, M. (2005) *False Alarm. The Truth About the Epidemic of Fear.* Hoboken, NJ: John Wiley & Sons, Inc.

Slocum, T., McMaster, R., Kessler, F. C., and Howard, H. H. (2009) *Thematic Cartography and Visualization* (3rd edn.). Upper Saddle River: Prentice Hall.

Sluga, G. (2005) What Is National Self-Determination? Nationality and Psychology During the Apogee of Nationalism. *Nations and Nationalism* 11(1): 1–20.

Smith, A. (2009) *The Internet's Role in Campaign 2008.* Washington, DC: Pew Internet and American Life Project.

Smith, A., and Rainie, L. (2008) *The Internet and the 2008 Election.* Washington, DC: Pew Internet and American Life.

Smith, C. D. (1987) Cartography in the Prehistoric Period in the Old World: Europe, the Middle East, and North Africa. In J. B. Harley and D. Woodward (eds.), *The History of Cartography Vol. 1: Cartography in Prehistoric, Ancient, and Medieval Europe and the Mediterranean* (pp. 54–102). Chicago: University of Chicago Press.

Smith, N. (1992) Real Wars, Theory Wars. *Progress in Human Geography* 16: 257–71.

Smith, N. (2003) *American Empire: Roosevelt's Geographer and the Prelude to Globalization.* Berkeley: University of California Press.

Snobelen, S. D. (2001) Of Stones, Men and Angels: The Competing Myth of Isabelle Duncan's Pre-Adamite Man (1860). *Studies in the History of Philosophy C: Biological and Medical Sciences* 32(1): 59–104.

Sparke, M. (1995) Between Demythologizing and Deconstructing the Map: Shawnadithit's New-Found-Land and the Alienation of Canada. *Cartographica* 32(1): 1–21.

Sparke, M. (1998) A Map That Roared and an Original Atlas: Canada, Cartography, and the Narration of Nation. *Annals of the Association of American Geographers* 88(3): 463–95.

Sparke, M. (2005) *In the Space of Theory: Postfoundational Geographies of the Nation-State.* Minneapolis: University of Minnesota Press.

Sparke, M. (2008) Political Geography – Political Geographies of Globalization III: Resistance. *Progress in Human Geography* 32(3): 423–40.

St. Martin, K., and Wing, J. (2007) The Discourse and Discipline of GIS. *Cartographica* 42(3): 235–48.

Stallman, R. (1999) The Gnu Operating System and the Free Software Movement. In C. DiBona, S. Ockman and M. Stone (eds.), *Open Sources: Voices from the Open Source Revolution* (pp. 53–70). Sebastopol, CA: O'Reilly Media Inc.

Stoller, M. (2007) What Is Openleft.Com? Retrieved February 23, 2008, from http://www.openleft.com/showDiary.do?diaryId=17.

Stoller, M. (2008) Dems Get New Tools, New Talent. *The Nation* 286(5): 20–4.

Stone, K. H. (1979) Geography's Wartime Service. *Annals of the Association of American Geographers* 69(1): 89–96.

Stone, M. (1998) Map or Be Mapped. *Whole Earth* 94(Fall): 54–5.

Surowiecki, J. (2004) *The Wisdom of Crowds. Why the Many Are Smarter Than the Few and How Collective Wisdom Shapes Business, Economies, Societies, and Nations.* New York City: Doubleday.

Surveillance Camera Players. (2006) *We Know You Are Watching: Surveillance Camera Players.* San Diego: Factory School.

Talen, E. (2000) Bottom-up GIS: A New Tool for Individual and Group Expression in Participatory Planning. *Journal of the American Planning Association* 66(3): 279–94.

Taylor, D. R. F. (ed.). (2005) *Cybercartography: Theory and Practice* (1st edn.). Amsterdam: Boston.

Taylor, P. (1990) Editorial Comment: Gks. *Political Geography Quarterly* 9: 211–12.

Taylor, P. J. (1992) Politics in Maps, Maps in Politics: A Tribute to Brian Harley. *Political Geography* 11(2): 127–9.

Teeters, N. K. (1957) *The Prison at Philadelphia, Cherry Hill. The Separate System of Penal Discipline 1829–1913.* New York: Temple University Publications by Columbia University Press.

Thompson, B. (2006) His Bottom Line: Educating the World's Kids. *The Washington Post,* September 9.

Thrift, N. (2006) *Non-Representational Theories.* London: Routledge.

Toledo Maya Cultural Council, and Toledo Alcaldes Association. (1997) *Maya Atlas.* Berkeley, CA: North Atlantic Books.

Tozzi, J. (2007) How Top Bloggers Earn Money. *Business Week,* July 13.

Turnbull, D. (1993) *Maps Are Territories: Science Is an Atlas: A Portfolio of Exhibits.* Chicago: University of Chicago Press.

Turnbull, D. (2003) *Masons, Tricksters and Cartographers: Comparative Studies in the Sociology of Scientific and Indigenous Knowledge.* London and New York: Routledge.

Turner, A. J. (2006) *Introduction to Neogeography.* Sebastopol, CA: O'Reilly Media Inc.

Tversky, A., and Kahneman, D. (1974) Judgment under Uncertainty: Heuristics and Biases. *Science* 185(September 27): 1124–31.

United Nations Development Program. (2006) *Human Development Report 2006. Beyond Scarcity: Power, Poverty and the Global Water Crisis.* Basingstoke and New York: Palgrave Macmillan.

United Nations Development Program. (2007) *Human Development Report 2007/2008. Fighting Climate Change: Human Solidarity in a Divided World.* New York: Palgrave Macmillan.

United States Joint Forces Command. (2007) *Geospatial Intelligence Support to Joint Operations.* Washington, DC.

Vasiliev, I., McAvoy, J., Freundschuh, S., Mark, D. M., and Theisen, G. D. (1990) What Is a Map? *Cartographic Journal* 27(2): 119–23.

Vinge, V. (2001) *True Names by Vernor Vinge and the Opening of the Cyberspace Frontier* (1st edn.). New York: Tor.

Vujakovic, P. (2002) From North–South to West Wing: Why the "Peters Phenomena" Will Simply Not Go Away. *The Cartographic Journal* 39: 177–9.

Wainer, H. (2003) Visual Revelations – a Graphical Legacy of Charles Joseph Minard: Two Jewels from the Past. *Chance* 16(1): 58–63.

Wainwright, J., and Bryan, J. (2009) Cartography, Territory, Property: Postcolonial Reflections on the Indigenous Counter-Mapping in Nicaragua and Belize. *Cultural Geographies* 16: 153–78.

Waldberg, P. (1997) *Surrealism.* London: Thames and Hudson.

Walker, F. A. (1874) *Statistical Atlas of the United States.* New York: J. Bien.

Walker, F. A. (1896) Restriction of Immigration. *The Atlantic Monthly* 77(464): 822–9.

Wallace, T. (2009, March 2009) Has Google Homogenized Our Landscape? Paper presented at the Association of American Geographers Annual Conference, Las Vegas, NV.

Wallis, H. M., and Robinson, A. H. (eds.). (1987) *Cartographical Innovations: An International Handbook of Mapping Terms to 1900.* London: Map Collector Publications for the International Cartographic Association.

Ward, R. D. (1922a) Some Thoughts on Immigration Restriction. *The Scientific Monthly* 15(4): 313–19.

Ward, R. D. (1922b) What Next in Immigration Legislation? *The Scientific Monthly* 15(6): 561–9.

Weber, S. (2004) *The Success of Open Source.* Cambridge, MA: Harvard University Press.

Wikipedia. (2007) Mashup. Retrieved June 26, 2007, from http://en.wikipedia.org/wiki/Mashup_%28web_application_hybrid%29.

Wilkinson, A. (2007) Remember This? A Project to Record Everything We Do in Life. *The New Yorker* 83(14): 38.

Wilkinson, S., Mackinder, H., and Lyde, L. W. (1915) Types of Political Frontiers: Discussion. *The Geographical Journal* 45(2): 139–45.

Wilson, L. S. (1949) Lessons from the Experience of the Map Information Section, OSS. *Geographical Review* 39(2): 298–310.

Winlow, H. (2006) Mapping Moral Geographies: W. Z. Ripley's Races of Europe and the United States. *Annals of the Association of American Geographers* 96(1): 119–41.

Winlow, H. (2009) Mapping Race and Ethnicity. In N. Thrift and R. Kitchen (eds.), *The International Encyclopedia of Human Geography.* Oxford: Elsevier.

Wood, D. (1992) *The Power of Maps.* New York: Guilford Press.

Wood, D. (2003) Cartography Is Dead (Thank God!). *Cartographic Perspectives* 45(45): 4–7.

Wood, D. (2006) Map Art. *Cartographic Perspectives* 53(Winter): 5–14.

Wood, D. (2007a) Lynch Debord About Two Psychogeographies, Retrieved August 15, 2007, from www.arika.org.uk/shadowedspaces/2007/lynch-debord/.

Wood, D. (2007b) A Map Is an Image Proclaiming Its Objective Neutrality: A Response to Denil. *Cartographic Perspectives* (56): 4–16.

Wood, D. (2008) The History of Map Art. *Counter Cartographies Convergence.* Talk given in Chapel Hill, NC, September 2008.

Wood, D., and Beck, R. J. (1994) *Home Rules.* Baltimore: Johns Hopkins University Press.

Wood, D., and Fels, J. (2009) *The Natures of Maps: Cartographic Constructions of the Natural World.* Chicago: University of Chicago Press.

Wood, D., and Krygier, J. (2009) Critical Cartography. In N. Thrift and R. Kitchen (eds.), *The International Encyclopedia of Human Geography.* New York and London: Elsevier.

Wood, P. (2003) Art of the Twentieth Century. In J. Gaiger (ed.), *Frameworks for Modern Art* (pp. 5–55). New Haven and London: Yale University Press.

Woodward, D. (ed.). (1987) *Art and Cartography: Six Historical Essays.* Chicago: University of Chicago Press.

Woodward, D. (1992a) A Devon Walk: The History of Cartography. In *A Celebration of the Life and Work of J. B. Harley, 1932–1991 [17 March 1992]* (pp. 13–15). London: Royal Geographical Society.

Woodward, D. (1992b) J. B. Harley (1932–1991). *Imago Mundi* 44: 120–5.

Woodward, D. (2001) Origin and History of the History of Cartography. In D. Woodward, C. D. Smith, and C. Yee (eds.), *Plantejaments I Objectius D'una Historia Universal de La Cartografia/Approaches and Challenges in a Worldwide History of Cartography* (pp. 23–9). Barcelona: Institut Cartografic de Catalunya.

Woodward, D. (2004) History of Cartography Project Broadsheet #12: The Map as Repository of Memory. Retrieved July, 2008, from http://www.geography.wisc.edu/histcart/broadsht/brdsht12c.html.

Woodward, D. (ed.). (2007) *Cartography in the European Renaissance* (Vol. 3). Chicago: University of Chicago Press.

Woodward, D., Smith, C. D., and Yee, C. (2001) *Plantejaments I Objectius D'una Historia Universal de La Cartografia/Approaches and Challenges in a Worldwide History of Cartography.* Barcelona: Institut Cartografic de Catalunya.

Wright, J. K. (1930) Density of Population of Belgium, Luxembourg, and the Netherlands. *Geographical Review* 20(1): 157–8.

Wright, J. K. (1942) Map Makers Are Human: Comments on the Subjective in Maps. *Geographical Review* 32: 527–44.

WSOCTV.com. (2006) Suspicious Liquid on Plane Identified as Water. Retrieved March 15, from http://www.wsoctv.com/news/9780677/detail.html.

Zetter, K. (2007) Eyes in the Skies Document Human Rights Violations in Burma. Retrieved October 1, 2007, from http://blog.wired.com/27bstroke6/2007/09/eyes-in-the-ski.html.

Zook, M. A. (2005) *The Geography of the Internet Industry: Venture Capital, Dot-Coms, and Local Knowledge.* Malden, MA: Blackwell Publishers.

Zook, M. A., and Dodge, M. (2009) Mapping Cyberspace. In N. Thrift and R. Kitchen (eds.), *The International Encyclopedia of Human Geography* (pp. 356–67). New York and London: Elsevier.

Zook, M. A., and Graham, M. (2007a) The Creative Reconstruction of the Internet: Google and the Privatization of Cyberspace and Digiplace. *Geoforum* 38(6): 1322–43.

Zook, M. A., and Graham, M. (2007b) Mapping Digiplace: Geocoded Internet Data and the Representation of Place. *Environment and Planning B: Planning and Design* 34(3): 466–82.

Index

Note: "n" after a page reference refers to a note on that page. Numbers in italic refer to an illustration or table.

LIBRARY, UNIVERSITY OF CHESTE.